Science and power

Also in this series:

The New Page
Memory of the Future

Science and power

Federico Mayor
Augusto Forti
with contributions by
Raymond Daudel, Franco Ferrarotti, Gérard Huber
and Alexander King
and a Preface by Ilya Prigogine
Edited by Nigel Hawkes

Challenges Series

UNESCO Publishing

The authors are responsible for the choice and the presentation of the facts contained in this book and for the opinions expressed therein, which are not necessarily those of UNESCO and do not commit the Organization.

Published in 1995 by the United Nations Educational,
Scientific and Cultural Organization,
7, place de Fontenoy, 75352 Paris 07 SP (France)
Layout: Jean-Francis Chériez
Composition: InfoPrint, Mauritius
Printed by Presses Universitaires de France, Vendôme

ISBN 92-3-103199-6

Preface

Ilya Prigogine

It is an unexpected honour for me to have been asked to write a preface for the book *Science and Power* by Federico Mayor and Augusto Forti. It is, however, a difficult, almost impossible task. The relations between science and power reflect the social organization of society. This is different in Western democracies from what it is in China; different again from the form it took in Ancient Egypt or in the Mayan civilization. It is an immense subject. Moreover, what do we mean by 'science'? What do we mean by 'power'? We have of course an intuitive feeling for what is involved. I agree completely with Federico Mayor that at the end of this century the relations between science and power are in a state of rapid change. All that I can hope to do in this preface is to describe some additional reasons for the situation we find ourselves in at the end of the century.

Never has the need to bring closer the relations between science and society, including its power structure, been so strong. Traditionally, science has been associated with our questioning of nature. It expresses our belief in the intelligibility of nature. In the words of A. N.

Whitehead, it expresses the goal 'to frame a coherent, logical, necessary system of general ideas in terms of which every element of our experience can be interpreted'.

But the very concept of nature is culturally bound. When I visited Japan for the first time in 1953, I had the privilege of meeting the great Japanese physicist, Hideki Yukawa. I was very interested to learn that in spite of the fact that he was a most successful scientist in the Western tradition, he did not accept the Western view of nature. He emphasized that in Chinese the word 'nature' means 'that which is so itself'. In contrast, Western science from Galileo to Einstein is based on the idea of the 'laws of nature'. Nature is in some sense 'obliged' to follow certain rules, which the human mind is able to discover and describe.

The Florentine architect and humanist of the Renaissance Leon Battista Alberti claimed that 'men can do all things, if they will'. But how can men do 'all things'? Francis Bacon had the answer: they have to obey the laws of nature. But to obey the laws of nature implies knowing what they are. Since then the ideas of omniscience and omnipotence have remained closely linked. Since then also, the temptation to extend this scientific model based on certainty to society has been great. August Comte wanted to found a 'social physics' modelled on astronomy, the stronghold of determinism. This was also the aim of John Stuart Mill. Power would have a rational basis; society would be part of the vision of a controlled, tamed nature. This however leads us directly to the nightmares described in the novels of Huxley, Kundera and Orwell, of a timeless society in which history has stopped, suppressed by violence and terror.

But today we have reached a new situation which implies a different image of nature and therefore also of science. We no longer believe in the image of the world as a huge piece of clockwork. We have come to the end of certitudes. At all levels of observation we have discovered a narrative element in nature. Whether in cosmology, in elementary particle physics, or in biology, everywhere we have discovered instabilities and fluctuations. Implicit in the classical view of nature was the idea that under well-defined conditions a system will follow a unique course, and that a slight change in the parameters will likewise produce a slight change in the results. We know now that this is only true for simplified, idealized situations. A radical change in perspective is occurring as a result of the realization that in general systems obey non-linear laws and display complex behaviour in the form of abrupt transitions, multiplicity of states, self-organization and unpredictability.

As the century comes to an end, we can see that the ideologies associated with science as well as with power are in crisis. Curiously, this brings the 'hard' sciences and the social sciences closer together. Indeed, the new conceptual framework offered by evolutionary complex systems brings to the social sciences a coherent set of ideas, such as multiple futures, choice and bifurcations, dependence on history, intrinsic uncertainty. These are important to our understanding of the challenges that face humanity over the next decades and bring new tools to the decision-makers. Examples are: in the environment, transport policy and technology assessment, which are still often based on short-term estimations using assumptions of equilibrium and optimization. We come back through these arguments to the ideas of 'social physics' and of a 'rational' foundation of power. But this extended physics and the rationality it involves take a new form, quite different to what we could have imagined even a few decades ago.

There are some other aspects that I would like to mention briefly. The amazing progress of science has already transformed the functioning of power. We are witnessing the start of an information-based multimedia society. But this stage leads also to a difficult choice. Never have there been so many potential applications to problems of direct importance to humanity, be they in the field of environment or in medicine. Therefore it may be understood that applied science is now at the forefront. But we should not forget that basic science, especially as it relates to the complex, unpredictable world we discover today, is still in its infancy. We need new mathematics and new experiments. Yet some eminent physicists unfortunately claim that fundamental science is near its end. There remain unsolved problems only in a few situations, such as very high energy (the Higgs boson, 'God's' particle) or the neurosciences. These views do not take into account the recent trends I have mentioned. To slow down the development of basic science today seems to me as irresponsible as would have been the repression of basic science at the time of Galileo, just before Newton's revolution.

A question often asked is 'What is the aim of science?' Of course nobody has the answer. But let me share with you my personal 'utopia', which shows that in spite of all the dramas of the twentieth century I remain an optimist. In all parts of the world, the transition to civilization has been going on simultaneously with the division of labour, caste systems and social inequality. To build the Egyptian Pyramids or the Great Wall of China, slaves were necessary. Violence has been an inherent part of civilization. Science enables us to entertain the utopia of a civilization in which violence is not obligatory, a world in which inequality is not a social necessity. The reduction of inequality both between and within nations is, on the whole, perfectly evident when we compare the situation today with that which existed at the beginning of this century. Here we see progress that could not have taken place without science, and we are back to the relation between science and power. It is, indeed, a fascinating subject and I am sure that this collection of interesting essays will play a catalytic role in stimulating further research. It is an area of fundamental importance for our future.

Contents

Introduction

Augusto Forti

The problems which are strictly linked to science and its applications have become an everyday experience, not only to scientists and politicians but to citizens everywhere. From genetic engineering, the source of a debate impinging on the fundamental ethical questions on which our society is based, to the fight against Ebola and HIV viruses; from the hope of eradicating hunger and poverty to the environmental catastrophes caused by pollution or to the natural ones caused by our inability to forecast – it is always science, its applications and its relations with power which are at the core of debate. The decisions that are taken are often hasty and sometimes wrong.

Science is a Pandora's box at which we gaze with a mixture of fear, hope and sometimes terror. But it has become one of the main components of our daily life.

This short volume tries to tell, through the stories of men, the history of culture and the personal experience of its authors, the difficult and sometimes tragic route which science has had to take in order to free itself from religious dogmas. Even when science was recognized as a fundamental component of our culture, the fight has continued, as science has been forced to defend itself from the embrace of ideologies or from the fatal attraction of politicians and power or economic lobbies.

Biology, in particular, has been used as an instrument of ideological exploitation. In *Mein Kampf*, Hitler declared that without a biological basis and a biological purpose, policy was totally blind. Nowadays we constantly refer to the concept of diversity, but we should not forget that this same principle was also upheld in National Socialist ideology, clearly summarized in the words 'the diversity of human beings has a biological basis'. The vital thing, of course, is not so much to recognize diversity as to accept both diversity and equality. But the concept of equality was contested by fascist ideology, founded as it was on the principle of fundamental inequality according to which some deserved to live and others did not. Indeed, the purpose of the recruitment of eugenists, psychiatrists, anthropologists and physicians by the state in Germany was to draw up a 'biological/hereditary document' for each citizen of the Nazi state, separating those considered deserving to live from those Germans who were Jews, gypsies or members of other minority groups.[1] Not to forget Lysenko's story in the Stalin era.

After the rise of modern science, symbolized by emblematic figures such as Bruno and Galileo, the Industrial Revolution transformed relations between science and power. In his book *The Counter-Revolution of Science*, Friedrich A. von Hayek sees the Industrial Revolution as the manifestation of a trend in which science and technology, but particularly science, became the reference for all human

1. See Gérard Huber's chapter on 'Ideological Deviation and Ethical Protection' on page 115 of this work.

knowledge and for any moral obligation. He says that the *esprit polytechnicien* occupies the whole human experience, so that all those experiences and attitudes, including those related to ethics, which are not susceptible to exact and scientific measurement, are deprived of any real value.

In his contribution to this book, Franco Ferrarotti seeks to put the Industrial Revolution into a wider cultural perspective, taking into account also the fundamental impact that the technical revolution played in the conceptual development of the *Encyclopédie* and in setting the basis of the Industrial Revolution. Capitalism, which rose from this important historical event, determined the emergence of a completely new society: new because it is based on a new concept of power and because science largely replaced the traditional values as a basis for moral obligations.

The confrontation between science and power today is the debate between science and scientists, jealous of their freedom, and governments which are the main financial source for research activities and often ready to 'abuse' science, even with the best of intentions. Abuse of power is not exclusive to governments; scientific lobbies, often allied with important industrial groups, can themselves subvert the process as has been the case with the atomic lobby which dominated research policy for nearly forty years.

The end of the Second World War is a turning-point which illustrates well this situation. It began the heavy involvement of governments in scientific programmes, outside the university area. Everybody remembers the Manhattan Project which, with the realization of the atomic bomb, brought the war with Japan to its end. Since then, we have seen a series of projects which, starting from a scientific theory and thanks to the joint efforts of scientists, engineers, industries and governments, were carried out with great expense and in a relatively short time, like the Apollo programme which landed man on the moon. This is the birth of 'big science' and today it is hard to conceive of an important scientific project which does not involve one or more governments.

Military research during the Second World War brought forward an incredible development of innovative technologies and products such as the jet engine, rockets to put satellites into orbit, radar, the development of electronics and computers, DDT, penicillin and many others. Science had been outstandingly successful in winning the war; it was essential that it should be equally effective in the peacetime economy and society. All over the world countries began to reassess their scientific resources with a view to reforming or creating a scientific infrastructure appropriate to the new situation. In the United States, where industry was so successful and so quick in exploiting new technologies, it was realized that many of the basic developments which had helped to win the war had originated from scientific research in Europe. It appeared necessary, therefore, to strengthen the basic research systems. President Roosevelt wrote to Vannevar Bush, the head of the exceptionally successful but temporary wartime organization, the OSRD, asking him to put forward suggestions for the post-war equivalent. This resulted in the historic Bush Report, *Science the Endless Frontier*, which eventually gave rise to the National Science Foundation.[2]

2. See the chapter, 'Science and Technology since the End of the Second World War', by Alexander King, on page 70 below.

As Federico Mayor remarks in his contribution to this book, we live in an age where science daily demonstrates its power. The knowledge it has given us has conquered distance, abolished want, reduced disease and opened doors to understanding the mysteries of nature. As an instrument for discovering 'the truth' – what 'lies behind' – science is without equal. Yet we should be wary of celebrating its triumphs too uncritically, for they have been unevenly distributed. In the midst of plenty, there is want; diseases long understood by doctors still carry off the poor and underprivileged; and in its tidy explanations of natural phenomena, science leaves many people unmoved or even alienated. What science tells us may be true, but it is not the only truth that matters.[3]

Heavy therefore is the responsibility of scientists. Unfortunately they have often been silent before the horrors which were taking place before their eyes. Scientists also have a responsibility towards our planet, towards the future generations and towards the underprivileged. The situation we face today is well summarized in the following words by the Nobel laureate, Dennis Gabor: 'Our present civilization is based materially on an extraordinarily successful technology and, spiritually, on practically nothing.'

More thought should be given to the social function of science and technology, recognizing that science and scientific knowledge are powerful tools for shaping our community. Among the most important is information, as Franco Ferrarotti rightly points out. Information has become the essential precondition of human development. During the

3. See the chapter by Federico Mayor, 'Science and Power Today and Tomorrow', on page 137 below.

eighteenth and nineteenth centuries, exploitation of man by man was direct and more or less precisely quantifiable. It had to do with manual labour, working hours, wage levels, and factory discipline. The scenario today is vastly different. The description of Volume 1 of *Das Kapital* is no longer tenable. The new variables are isolation, segregation, solitude, neglect and exclusion. To be exploited in the modern world means to be on the fringes, to be cut off. Power today exploits and oppresses not by the use of direct action, but simply by ignoring, by failing to intervene, by refusing to take action, by taking refuge behind complex and perfectionist procedures through which legal formalism and political paralysis support each other.[4]

But in order *to participate* today we need more education and literacy in science, spread more widely among politicians and journalists. They should be able to present scientific facts, which are at the base of important decisions, in a plain and comprehensible way so that all the citizens can be part of the decision-making process.

At the beginning of this century, it was widely accepted that power would have a rational basis. This brought as an extreme consequence the scientification and rationalization of society of which I spoke above, and the deterministic approach which brought us to the nightmares of a timeless society very well described in the novels of Huxley, Kundera and Orwell.[5] But a turning-point appeared in the second decade of our century. Since the work of Heisenberg and others, we have witnessed the

4. See the chapter by Franco Ferrarotti, 'The Industrial Revolution and the New Assets of Science, Technology and Power', on page 47 below.
5. See Ilya Prigogine's Preface to this work.

rise of a new science which gives us a completely new vision of the world.[6] Today we no longer believe in the image of the world as a huge piece of clockwork. We have a different image of nature, and therefore also of science. We have come to the end of certitudes. At all levels of observation we have discovered a narrative element in nature. Be it in cosmology, in elementary particle physics, or in biology, everywhere we have discovered instabilities and fluctuations. Implicit in the classical view of nature was the idea that under well-defined conditions a system will follow a unique course, and that a slight change in the parameters will likewise produce a slight change in the results. We know now that this is only true for simplified, idealized situations. A radical change in perspective is occurring as a result of the realization that systems obey in general non-linear laws and display complex behaviour in the form of abrupt transitions, multiplicity of states, self-organization and unpredictability. So, as the century ends, we can see that the ideologies associated with science as well as with power are in crisis.[7]

The time is now ripe to reconsider the relationship between science and technology, on one hand, and political power and government on the other. In particular, governmental science policies can no longer be decided on the basis of expertise provided by a small team of experts, but should instead be the result of a lengthy process in which all elements of society affected by the decisions should be involved. This is what we might call 'returning science to the citizens'. In addition, the content and

6. See Augusto Forti, 'La morte di Newton', *Prometheus* (Franco Angeli, Milan), 1990.
7. See the Preface to this work by Ilya Prigogine.

objective of international programmes in science should be reoriented to better serve basic human needs, the whole of humanity and not only the privileged few.[8]

Of course this new vision will require new attitudes. Politicians should be more concerned with the medium term, rather than making short-term decisions often biased by a demagogic flavour and sometimes with negative and irreversible effects on the environment. They should be ready to cope with the instability of the world and the uncertainty of the advice received. Scientists, too, should have a greater sensitivity towards the social and political impact of their advice. But as Federico Mayor says, this indispensable interaction between science and politicians can flourish only in free and democratic societies.

We have presented here some thoughts about the relations between science and Western society. But there are other cultures, and the relation between them and science should also be analysed. In a utopian vision, we could imagine science playing a fundamental role in creating a more tolerant world and as an indispensable element in building a culture of peace. The relations between science and power, as we have seen, are never static. They evolve continuously and we hope that this book will contribute towards stimulating the debate on this crucial issue.

8. See Augusto Forti and Paolo Bisogno, *Research and Human Needs*, Oxford, Pergamon Press, 1981, and Federico Mayor, *Scientific Research and Social Goals: Towards a New Development Model*, Oxford, Pergamon Press, 1982.

Knowledge and science as a source of power: the birth of modern science

Science, philosophy and power in Classical Antiquity

Augusto Forti

Classical Antiquity and the Greek and Roman worlds are the source and foundation of modern science. Without Democritus, Aristotle, Pythagoras, Ptolemy, Archimedes, Lucretius, Vitruvius and others, we would not have had Newton, Kepler, Galileo or Einstein. Ariadne's thread thus ran its course, coloured by science, technology and philosophy, from Athens to Alexandria, then to Rome, and from Heidelberg to the Pasteur Institute in Paris, culminating at Los Alamos, New Mexico. Along the way it revealed to us the passionate story of the world's scientists, today masters and tomorrow victims of power – giants like Pythagoras, Galileo and Sakharov, and occasional monsters such as Lysenko, the physicians at Auschwitz, the various Doctor Strangeloves.

But the key to understanding is found by tracing back to the Graeco-Alexandrine-Roman culture. For a multitude of reasons, a dazzling well-spring of observations, hypotheses, theories, and discoveries burst forth in Greece and the Mediterranean, and came to be the origin of modern scientific thought. It was during this period that Western science was born and man started to elaborate more complete scientific and cosmogenical theories. And it was in this environment, nourished by Oriental and Mediterranean culture, that man first began to ask himself rational questions about his position and role in relation to nature and the universe. This was also the genesis of all the problems of our contemporary society, including the relation between science and power and between science and technology.

It was a scientist who was the first to identify common features between political order and natural order. He analysed political power using a methodology that aspired to the same rigorous objectivity as that applied to natural sciences. This was Aristotle (384–322 B.C.), pupil of Plato and preceptor of Alexander the Great, founder of the Peripatetic School. Every object is defined by its immediate genus and by the qualities of its particular species. In his definition, man is a political animal. 'Animal' represents the genus and 'political' represents the difference determining the species.

Aristotle provides us with the first classification of power and government, according to the various ways through which power is used and distributed. According to him, governments can be categorized into six different forms:

> The government is the supreme authority of the state and is naturally in the hands of one person, or a small group of people, or the mass of citizens. When the power lies with one person, or a small number, or the majority of the people, and is exercised in the common interest, institutions are just and valid. If the power is exercised in the sole interest of one person, or a small group, or the masses, then these represent deviations of the above-mentioned cases. If

the former cases of government are defined in terms such as monarchy, aristocracy and republic, then the corresponding degenerations would be tyranny, oligarchy and democracy (abusive interest of the many on the few).

The last period of ancient science, which is called the Graeco-Roman period, starts with the Roman conquest of Egypt and ends up in the fourth century A.D. The decay of the city-states, which were centres of thought, action, scientific research and technological innovation, marked a turning-point in the relationship between science, society and philosophical thinking. This turning-point coincides with the development of autocratic oppressive power culminating in the deification of Caesar. In reaction to the absolutist state, scientists like Epicurus and Lucretius, realizing their impotence in the face of the power of the state, turned back in on themselves, seeking a moral perfection independent of the outside world and therefore of temporal power. Others, like the Stoics, while still analysing the internal self, were ready to put their findings at the service of power, so signalling the end of the great utopian visions such as those of Pythagoras and Plato. We can underline the ethical precepts already in existence at the time of the Sophists and Democritus by examining the thinking of various individuals.

Epicurus was without doubt one of the outstanding personalities of the period. Born at Samos in 342 B.C., he had as teachers Plato and Democritus. Of all thinkers of ancient times, Epicurus was the one who gave science the foremost position in the frame of culture. He understood that science needed to provide a vision of the world from which man could take inspiration in order to free society from superstition and tyranny.

Epicurus remarked that science is important because only science can give us happiness. The study of nature is not intended merely to allow a man to proclaim and demonstrate his knowledge before his fellow man, but on the contrary to produce serious and independent individuals, able to appreciate true, personal qualities and not just the outward appearance. In his letter to Pythocles, Epicurus says that being able to explain a phenomenon is more important than its actual occurrence. The order of the celestial phenomena must be explained by analogy to the natural phenomena that occur on earth, and recourse should never be made to 'divine nature' as an explanation.

He established his school at Mitylene, and, following the Pythagorean tradition, accepted women and slaves. For this reason, Cicero described the Epicureans as 'plebeian philosophers'. Epicureanism became, in fact, the doctrine of the people, in contrast to Stoicism which in both Rhodes and Rome had become the philosophy of the privileged class.

This set the scene for the first serious clash in history between science and constitutional power. The movement which he initiated spread widely throughout the entire Hellenistic world, reaching as far as Rome. By 170 B.C., it had achieved such social and political importance that in that year the Roman Senate had to chase off two pupils of Epicurus: Alceus and Philiscus.

The Epicureans were using science in their attempt to reform theology, trying to separate religion from natural laws in order to have a better defence against authoritarian power. Under their intended reforms, the deities would begin to lose their grip on this world and become only a transcendent factor of knowledge and happiness, informing the consciences of wise men. This vision of the gods implied also that they were not responsible for the happenings in the physical universe and in man's social life. The human soul was made up of atoms and disintegrated at death with the body. Immortality did not exist.

At that time in Rome, an intense struggle for power was going on between conservatives and those who wanted to bring about reforms and social innovations, which included the Epicureans and also the Hellenizers, the so-called circle of Scipios and the *populares* who were driven by Gaius and Tiberius Gracchus. Epicurus, by raising the problem of living by following nature as one of the basic tenets of his doctrine, naturally clashed with the many soothsayers, priests and priestesses who served the official Roman power, and began indeed to threaten this power. The Epicurean movement thus became an element of social revolt against those in power and a promoter of new laws of reform, similar to the new agrarian laws which had been proposed by the Gracchi.

This new vision of the physical world, which came in great part from the ideas of Democritus, was the weapon with which an enlightened minority tried, unfortunately in vain, to dismantle the power structure of that time. The Epicureans inevitably clashed with the leading Roman class, which was maintaining its power by using the state religion, so dear to Plato, and superstition, and which was later to introduce new forms of superstition, such as the cult of the divinity of the emperor, a concept imported from the Hellenistic kingdoms.

Polybius, who was taken as a prisoner to Rome and who was soon to become a member of the Scipio circle, wrote that the true cornerstone of Roman power was superstition. Of course we know that this is not the whole truth because the bedrock of Roman power was not only social stability maintained through an autocratic regime, but also the development of technology, which in turn had a strong influence on Roman expansion (roads, communication, and steel for weapons and war machines).

Epicurus' fundamental role in the relationship between science and society was nourished by the scientific and social theories of Democritus and other prominent scientists. The atomistic intuition of Democritus represented, in fact, the highest vision of the natural system governing our physical world. Epicurus took to the limit his theories about the importance of science as a fundamental aspect of culture, and therefore as an active element in social conflicts and in the struggle for power. Always drawing inspiration from Democritus, Epicurus and, to an even greater extent, Lucretius took a fundamental step in defence of the individual's freedom and the freedom of his will. Democritus asserted that, instead of moving in all directions, atoms fall from top to bottom under the action of their own weight. But in order to produce the collisions that create bodies and entities from the individual atoms, one has to accept the idea that the atoms may suddenly deviate from their path. This is the theory of 'Parenclisis' produced by Lucretius.

We do not know whether Epicurus was the first to propose this theory, partly because almost all his works have been destroyed by accident or by the bitter jealousy of his rivals. But he wrote to Meneceus:

> Our will is autonomous and independent and to it we can attribute praise or disapproval. Thus, in order to keep our freedom, it would have been better to

remain attached to the belief of goodness rather than being slaves to the fate of the physicists: the first gives us the hope of winning the benevolence of deities through praise and sacrifices, the latter, on the contrary, brings with it an inviolable necessity.

If the soul is made of atoms, he thought, and these are governed by very strict physical laws, there is no space left for free will. We do not know what Epicurus himself thought about this dilemma, but Lucretius, who better than anybody else is the interpreter of his thinking, finds in the physical laws themselves the solution to individual freedom and free will. In some ways this is an anticipation of the principle of indeterminacy defined by Heisenberg many centuries later.

As Lucretius put it:

If we have to suppose a connection between all the motions and, because of a general law, that a motion is always born from a preceding one and, if the primordial atoms should deviate, would not then a new situation result which breaks the law of destiny? If this were not to happen, how could human beings break a predetermined destiny, and go and act where and as their mind wishes?

Another important figure among the scientists and thinkers of this period is Zeno (336–264 B.C.), a contemporary of Epicurus known for his famous mathematical paradox and founder of the School of the Stoics, which was to become so important in the struggle for power in Rome. The original Stoa preached unity between city and state (the republic of Zeno), saying that only a man who conforms himself to natural law could bear the title of citizen of the world. But the Stoics still believed in esoteric prediction and in the divinity of the heavenly bodies, accepting astrology and prediction. Revolutionary ideas were also common, for example the influence which the Stoic Sphaerus had on the socialist experiment of King Cleomenes from Sparta. The idea was that if men are part of a unique community and equal in their rights as a result of natural law, why should they not also participate on an equal basis in the enjoyment of material goods? The cosmopolitan attitude of the Stoics, accepting the principle of divinity, fitted in very well with the development of power in Rome, and their ideas became well established.

The Stoics supported the idea of the immortality of the soul. Their cult of divinity, borrowed from Oriental culture, gave better protection to the Roman oligarchy, when threatened by social conflict, than was possible with the weak local or Etruscan deities. Cicero, using their arguments, was able to defend the commercial spirit and the privileges of oligarchic power.

There have been many discussions about why science decayed in the Roman Empire, to the advantage of technology. Romans were mainly warriors and country people, and in these two areas they made great strides, whilst their contribution to science was meagre and disproportionate to their political importance. They were builders on a major scale, producing edifices, aqueducts and roads, while medicine also made tremendous advances. But very little indulgence was given to scientific explanations of natural phenomena. Such explanations were seen as a threat, because anyone attempting them challenged the powers of the divinities and astrologers who served the ruling class. This was the main reason for the slow development of science. It needed the courage of a forerunner to Galileo or Bruno like Lucretius.

Rome produced physicians like Celsus and Galen, geographers such as Polybius and Strabo, architects like Vitruvius, and agriculturists like Cato and Pliny. But this is a long way from the scope and ambition of Greek and Alexandrine science, and one might be tempted to conclude that the development of science is linked to freedom and social changes.

References

FORTI, U. *Storia della scienza.* Milan, Dall'Oglio Editore, 1969.

——. *Tecnica e progresso umano.* Milan, Fabbri Editore, 1963.

TATON, R. 'Histoire générale des sciences' series, Paris, Presses Universitaires de France, 1983.

The birth of modern science and freedom of thought

Augusto Forti

After the fall of the Roman Empire, the ideas that underpinned the rise of science in the seventeenth century were mushrooming in Byzantium and the Arab world. Even under the barbarians, great scientists like Boethius were active at the court of Teodorico in Verona, teaching and carrying out research. Sometimes, as with Boethius himself, they fell out of favour with the Prince and died under torture because of the jealousy and envy of the courtesans.

But the birth of modern science, it can be confidently stated, coincides with the work of Galileo and the triumph of empirical methods. The development of science and mathematics in the seventeenth century reached a level unknown in previous eras. This was a result of the cultural revolution of the Renaissance and of the development of trade and economics linked to technological progress. It is no wonder that this century has been named the 'century of genius', when an array of scientists such as Galileo, Torricelli, Descartes, Pascal, Malpighi, Huygens, Newton and Leibnitz can be set beside their contemporaries, the great masters of the arts, Shakespeare, Milton, Spinoza, Molière,

Calderón, Gronzio, Rubens, Rembrandt, Velázquez, Caravaggio and Palladio, to name just a few.

What created the conditions for such a burgeoning of the human genius? In particular, the birth of modern science is due to the changes in the social structure at the end of the Middle Ages. Craftsmen and budding innovators were no longer a class considered inferior because they possessed manual skills, as had been the case in ancient Greece. Instead, they became a class increasingly independent of the traditional powers. European society, entrepreneurial and commercial, was ready to nourish the dynamic and innovatory turn that modern science was taking. Leading the way was the technological revolution of earlier centuries, which made possible a new scientific vision of the world.

It is now that one of the major figures in the history of science emerges, one who was to develop his theories and carry out research that has made such a profound mark on the history of science and indeed of humanity: Galileo Galilei (1564–1642). He is also an emblematic figure in our portrait gallery of the greatest actors on the science and power stage. But let us place this story in the proper cultural context.

The value of the method of induction was becoming increasingly recognized, and Galileo's work marked not only the flowering of Renaissance technology, but also the beginning of modern science. Galileo was responsible for three major concepts in the history of ideas. First, nature is rich with facts that conform to precise laws and, as astronomy shows, even a mechanism of gigantic dimensions can be interpreted on the basis of laws deduced from the behaviour of the commonest bodies, such as a projectile. For this reason, our intellect can comprehend the inner truth of natural events. Finally, given that this truth is expressed by laws that are essentially mathematical, calculation just as much as geometry constitutes the ideal model for reason. If philosophy is to escape from the shallows of generality, it too has to recognize this truth.

Galileo and his successors thought of modern science as a mathematical tool for discovering the absolute truth of nature. Complexity is only apparent, and can be broken down into a series of simple and universal laws describable in purely mathematical terms. For Galileo, certainly not an atheist, the soul created in the image of God is able to penetrate the truths that govern creation, and in so doing progress towards a knowledge of the world, which God possessed in full at the beginning. And the truths discovered in nature by science are truths of divine creation and thus carry validity, even if they are in apparent contradiction to the Holy Scriptures. It was this very argument, already supported by Bruno and Campanella, that angered the theologians, who until that time had been the depositaries and only interpreters of the truths revealed by the Holy Scriptures and consequently the holders of enormous power.

In 1609, Galileo built his own telescope and used it to reveal a new cosmological reality, with the observation of the satellites of Jupiter and of the phases of the moon. Scientists and philosophers such as Telesio, Bruno and Campanella had already foreseen this new reality, with the critics of the Scholastic and the Aristotelian philosophers.

The publication of Galileo's work *Siderius nuncius* (*Celestial Messenger*), which was basically Copernican in all its implications, gave rise to a harsh confrontation between the new ideas in science, the traditional philosophers, and the Church. Contrary to Thomas Scott or William of Occam and other thinkers (theory of the double truth),

Galileo bluntly affirmed that the idea of two truths is a nonsense. In a period when the Inquisition was becoming more and more belligerent, Galileo had the courage to state that the Holy Scriptures often need an explanation that is different from the words, and that in any case in the controversy concerning natural laws the Bible should really play a secondary role.

At the same time he was defending the rights of human ingenuity and imagination. Who would dare to assert that we already know everything that is to be known and discovered in the world? Biblical interpretations have to adapt to the conclusions of science – the result of research, experiments and demonstrations – and not vice versa. This was the revolutionary view that Galileo supported and for this alone we can say that we owe him the birth of modern science, which stands alone in its own environment and inspiration.

The revolutionary ideas of Copernicus and Galileo not only gave rise to harsh confrontations between science and religion, but also to conflict within the Church of Rome itself. A sharp battle developed between the Dominicans, who at that time represented the conservative and reactionary wing of the Catholic Church, and the Jesuits who were more open to culture and science. At the beginning of the seventeenth century, the Jesuit mathematicians and astronomers of the Collegio Romano were recognized among the highest scientific authorities.

But the publication in 1632 by Galileo of the *Dialogo dei massimi sistemi* (*Dialogue on Two World Systems*) made the position of Pope Urbano VII, who at the beginning had been one of Galileo's indulgent supporters, much more difficult. That book, the late acceptance by the Protestants of the Copernican theories, and the fight that the Jesuits were conducting against the Reformation, made Galileo's trial and condemnation by the Inquisition inevitable. So one of the world's greatest spirits, nearly blind and seriously ill, ended his days in the hands of the Holy Inquisition on 8 January 1642. Galileo died but, after a very dark period, science finally won its battle against the Church for freedom of thought.

Other scientists such as Lavoisier, during the French Revolution, or those who opposed the ideas of Lysenko in Stalin's Russia, paid with their lives the price of expressing their scientific theories and ideas against the power of the time. But after Galileo and the birth of modern science, we can say that the battle between science and power gave rise to the new vision of the world where science became a fundamental, strong and independent element in the history of mankind.

Galileo's views did not, of course, emerge from nowhere. To understand them we must step backwards to Copernicus (1473–1543) who was the greatest scientific personality of the end of the Middle Ages. His cosmology was one of the main driving forces of the Renaissance and the source of a hard struggle between science on one side and religion and its temporal power on the other. Copernicus grew up and studied in Poland and Italy, at the universities of Torum, Bologna, Padua and Ferrara. Today, we might see him as an emblematic figure for a Europe seeking unity in order to provide for its own political and cultural future. His new geometric syntax had a fundamental influence on the evolution of the scientific and philosophical ideas of Bruno, Galileo, Descartes, Newton and Voltaire. But it was only fifty years after Bruno's and Galileo's argument with the repressive forces of traditional philosophy and religious dogma that, thanks to Newton's *Principia*, these new geometrical and astronomical ideas reached the apotheosis in the most vast and prodigious rational construction of all time.

Despite the innovative atmosphere of the Renaissance, all the established cultural forces were set against Copernican theories in the second half of the seventeenth century and up to the first decades of the eighteenth century. Traditional Aristotelian and scholastic philosophers were opposed not only because the new ideas appeared too difficult from a mathematical viewpoint, but principally because they felt it necessary to continue to keep the earth at the centre of a well-organized universe, the same universe described by Dante in his *Divine Comedy*. They instinctively felt the danger to their own position from the new model. At that time, physics was continuing to ignore the principle of inertia outlined by Galileo, and therefore philosophers could discuss the heliocentric theory as scientifically unacceptable. Finally, the power of religious dogmatism, embodied by personalities like Calvin, Luther, Bellarmino and Urbano VII, regarded these theories as heretical. In 1541, for example, Melanchton asked the authorities to take energetic action against these 'irresponsible fantasies', and Luther treated Copernicus as a crazy man. Pope Clement VII, by contrast, looked with great admiration at Copernicus' work, but by the end of the century the Church had discovered that a snake haunted this paradise of geometrical splendours.

Of the Copernicans, Bruno and Galileo were the most eminent. Galileo, as we have seen, originated a new conception of mechanics and astronomy which established the basis of modern science. Bruno, on the other side, gave birth to a new pantheist vision of the world and of life, which were then developed by Campanella, Spinoza and many other romantic thinkers and writers. Copernicus himself died in May 1543, hiding his last works under his bed.

Galileo's contemporary Sir Francis Bacon (1561–1626) was another fundamental figure in the renewal of science and its new relation to society. A prolific writer and empirical scientist, he died while testing whether meat would be preserved if left out in the snow. In his work he was fond of describing a utopian society similar to that of Sir Thomas More's *Utopia* or Campanella's *City of the Sun*. But his main contribution to the building of a new society was his intuition that technology would play a fundamental role in creating the welfare for mankind.

At that time, two parallel schools of thought had emerged to modify the balance of power in favour of science. One school was that of Bruno and Galileo, and therefore more closer to philosophical principles, while the other was that of Sir Francis Bacon who already envisaged the role of technology in support of the renewal of society. His theories were an important ideological contribution to the illuministic revolution of Voltaire and Dalembert. During the restoration after the French Revolution, his ideas were once again attacked, particularly by Demestre, as utopian nonsense.

Europe in the second half of the sixteenth century was home, as we said above, to another free spirit who contributed to the progress of scientific thinking. To Giordano Bruno (1548–1600) we owe the concepts of relativity and infinity; he believed that the universe was infinite, and filled with innumerable planetary systems containing life. He asserted also that absolute truth did not exist, and that one's perception of the world depended upon one's position in time and space: two conceptions that a modern scientist would find no difficulty with, but which brought him into conflict with the Church. He was imprisoned by the Inquisition in 1591, tried, and burned at the stake in 1600.

When in Prague, he wrote the following prophetic words:

> I do not like to hide the truth which I see, nor am I afraid to openly profess it. As everywhere I have been involved in the fight between darkness and light, between science and ignorance, therefore everywhere I have been the victim of hate and I have experienced the anger of the dull and stupid multitude as well as that of the official academics who are the fathers of ignorance.

Even today a free spirit would share these sentiments when faced with intolerance.

A generation later, another major figure emerged – René Descartes (1596–1650), who is considered the master of modern thinking and a fundamental influence on the Illuministic movement. He was a physician by training, but made many contributions in the field of physics and mathematics. He preached full acceptance of Copernican theories and presented a hypothesis of the origin of the earth which was in many instances in contradiction with Genesis. His book *Le Monde: ou le traité de la lumière* had to wait many years to be published because Descartes was advised by his friends of the danger that its publication could represent for his life. By this time, scientists were beginning to convene more to exchange ideas and confront theories. Descartes himself was a skilled swordsman, who crossed Europe, fighting with the Bavarian troops then in the service of the Duke of Savoy, and he was also present at the siege of the Huguenots at La Rochelle. This period saw the beginning of the Royal Society, created in 1662, and the French Royal Academy for Science, in 1666. Thus began the 'Invisible College' and the international spirit of tolerance against the dogmatic impositions which science had to tolerate for many centuries. Thomas Sprat, who was one of the founders of the Royal Society, said that 'the natural philosophy does not split up into factions but it permits our ideas to differ with no animosity and allows antagonistic points of view to be confronted without the danger of a civil war'.

In 1633, Descartes's scientific philosophical treaty *Le Monde: ou le traité de la lumière* was completed. In it he accepted the Copernican and Galilean ideas on the movement of the earth, the concept of an infinite universe, and many other points in conflict with Genesis. When he heard the news of Galileo's condemnation, he put the manuscript aside and the book was published only fourteen years after his death. The conflict between scientists and the religious power was still acute, to the point where the Church could include in the Index of banned works the whole of Descartes's output. When his remains were brought back to France from Sweden (where he died at the service of Queen Christina, it is said because of the extreme cold both in the royal palace and in the library), the mathematician Jacobi commented that, after all, it was much better to have the ashes of a great man than to have him alive.

Descartes was more inclined to mathematics than physics and his *Discours sur la méthode* is another cornerstone in the birth of modern science. From his '*cogito ergo sum*' and from his remarks on the primacy of mathematics, the various elements of the puzzle have brought us step by step to the birth of modern scientific, philosophical and humanistic culture. Its inevitable consequence was the redemption of science from the power of the Church and the dogmatic beliefs which were still very strong at that time.

Spinoza (1632–77), another free spirit in the Europe of the second half of the seventeenth century, brings Cartesian thinking to its culmination. In his *Etica more geometrico demonstrata*, written with definitions derived from geometrical and mathematical concepts, he supported the idea that thinking in an honest way means elaborating one's own thoughts and matching them with reality, because the only correct way for philosophy is the one that follows the mathematical approach for which Euclid provided the model. He also stated that the Church should be subordinated to the state, and in his *Tractatum theologico politicus* declared that citizens should be granted complete freedom of expression.

As we said above, the birth of modern science had also a fundamental role in the internal debate for the renewal of the Church. In this context, we cannot overlook the mathematician and physicist Blaise Pascal (1623–62). He was involved in a dispute that concerned Jansenists and Jesuits, and the Abbey of Port Royal des Champs. Jansenius (1585–1638), the Bishop of Ypres, had preached in his book on St Augustine for the freedom of religious thought against the formalism and the rigidity of ecclesiastic hierarchies. Thus the Jansenists were ready to defend freedom of thought against the Jesuits, who were in the service of authority and accused of moral laxism and opportunism. The Jansenists made the Abbey of Port Royal one of the most lively and open centres of the religious renewal. They were open to new scientific ideas and adopted Cartesian logic. As Pascal recorded in his *Thoughts*, Cardinals Mazarin and Richelieu opposed this centre of free spirits, which they deemed insufficiently papist and insufficiently royalist. Louis XIV, who was strongly influenced by his Jesuit advisers and who was already known for the massacres of Protestants, destroyed this old

Cistercian convent, violated the graves and scattered the remains. In spite of that, the Jansenistic spirit of the centre lives on in Pascal's work. He rejected the *esprit géométrique* which was so common at that time, and in contrast to Descartes, asserted that rigorous mathematical thought can be used in order to describe experience, but cannot be used to prejudge the results or to constrain them within rigid schemes. He explored, in this way, the new mysterious frontiers of mathematics, which were put aside by Descartes, such as the theory of indivisible numbers and the new conceptions of the infinitesimal. Pascal's thoughts and intuitions are incredibly modern, as illustrated in some chapters of his book *De l'esprit géométrique* in which he anticipates the Popperian theories of falsification.

Like a jigsaw puzzle, with Copernicus, Galileo, Bruno, Descartes and all the other great thinkers of this historical period, piece after piece the great fresco of a period in which were laid the roots of a modern society is completed. Science has become a fundamental element of our culture and human dignity.

We cannot conclude this overview on the birth of modern science and its relation with power without remembering the fundamental role of Isaac Newton (1642–1727) who was born in the year Galileo died. While still young, he obtained the chair at Cambridge after he presented a paper of deep interest in the field of optics and mathematics. With his mathematical solution of the elliptical orbits of the planets and the computation of the masses of the sun and the planets, he was able to open for ever the way to the new cosmological theories. The mathematical system built by Newton was the most extraordinary product the human mind has ever been able to realize. This was a very positive time for science.

The new cosmology was progressively accepted, while James II, who was favourable to the Jesuits and a friend of Louis XIV, was removed from the English throne and replaced by the liberal William of Orange (1688).

This period, known in England as the 'Glorious Revolution' and marked by a new pact between the king and the people (Declaration of Rights), saw the birth of the freedom of the press and the abolition of trial by torture. Newton participated in the construction of the new state and became a Member of Parliament for Cambridge. He was also secretary of the Royal Society and later General Controller of the Royal Mint.

We can say, therefore, that at this period the conflict between science and religious dogma, which required such a high tribute of victims, was finally resolved. Science was finally free from all dependency and from temporal and religious power. With the birth of modern science and, at the same time, modern thought a new era started. The battle for freedom of thought was not finally won, of course. In the future, authoritarian regimes would claim victims among the scientists who resisted their ideologies. But the power of religious belief to stand in the way of science was broken, never to regain its obstructive potential.

The Industrial Revolution: science, technology and power

The Industrial Revolution and the new assets of science, technology and power

Franco Ferrarotti

The concept and the term 'Industrial Revolution' are far from being universally accepted. Historians and economists of a conservative disposition tend to criticize them to a greater or lesser extent. But my view is, primarily, that the Industrial Revolution was a true revolution, that is, an epoch-making break with the past and, second, that it has not so far been fully analysed in its impact on global society. The Industrial Revolution has of course been explored and discussed, especially by economic historians in its social, economic and political aspects. This is necessary but not sufficient. The critical examination of its broad cultural and scientific impact has lagged behind. In particular, it has not yet been fully recognized that power and science, as a result of the Industrial Revolution, changed their nature. The basis of legitimacy for power has shifted from traditional or democratic acceptance to rational accountability. Science has become a group enterprise and an organizational weapon capable of deeply influencing the political power structure, the economic system of production and the overall social and intellectual climate. In this respect, I submit that the Industrial Revolution was a genuine revolution, a conception which is essentially alien to classical political thought.

The most vocal critic of the concept of the Industrial Revolution has been the well-known economist Friedrich A. von Hayek.[1] Leaving aside polemical exchanges, it is safe to say that Hayek's general theoretical position is best expressed in *The Counter-Revolution of Science*. The Industrial Revolution is here seen as a manifestation of a more general trend in which science becomes the sole source not only of human knowledge but also of any moral obligation. In this way, the triumph of the *esprit polytechnicien*, as he calls it, covers and comprises the whole of human experience in such a way as to deny any value, both cognitive and ethical, to attitudes and experiences not susceptible to exact, 'scientific' measurement. If we follow Hayek's logic, science is reduced to 'scientism', leading necessarily to a dogmatic position, which constitutes a total denial of a truly human problematic awareness, and therefore of science itself.[2]

In the judgement of T. S. Ashton, the concept of the Industrial Revolution was a product of revolutionary agitators in the second half of the eighteenth century, certainly not the result of a proper historical

1. See F. A. von Hayek (ed.), *Capitalism and the Historians*, Chicago, University of Chicago Press, 1954.
2. It is hardly necessary to recall that the most comprehensive criticism of 'scientism' of Galilean ascendancy is to be found in the works of one of the founders of phenomenology, Edmund Husserl.

investigation. It was used and spread extensively by French revolutionary orators, and not by British economists or industrialists. It is one of those catch-phrases that function like doctrinaire beliefs or effective ideological barriers, preventing the scrupulous historian from performing his duty properly.[3]

From this peculiar point of view, it is no wonder that Ashton denounces the abstract or doctrinaire nature of many theories, even when propounded by such classical authors as Marx, Sombart or Schumpeter. Ashton's critique is powerfully expressed, but the historian has a duty to go beyond petty details in order to achieve a meaningful synthesis, not be content with a minute, painstaking philological exercise. History is, and must be, a coherent interpretative narrative and not simply a sequence of discrete, albeit scientifically ascertained, facts. On the other hand, the good historian should refrain from applying a purely formal and, to an extent, arbitrary dialectical scheme to a single historical phenomenon. If the historian gives in to such external dialectics, history not only loses its human, dramatic and peculiarly unpredictable character. Its meaning is actually lost. It becomes simply the gradual emergence of the inevitable.

What seems to escape the participants in the debate about the nature of the Industrial Revolution, as edited by von Hayek in *Capitalism and the Historians*, is the fact that far from being confined to an internal dispute between historians, the real issue concerns the essence of capitalism,

3. For a general examination of the position of T. S. Ashton, especially expounded in *The Industrial Revolution*, see F. Ferrarotti, *Il rapporto sociale nell'impresa moderna*, (Rome, Arnaudo Ed., 1961), in particular pp. 15–21.

its relation with science and the broad consequences of its advent in Europe during the eighteenth century.

The question of the transition from the Middle Ages and the feudal social system to the modern capitalist system has often been squarely faced but usually in a narrow-gauge perspective which was bound to miss the interconnection between its various aspects, in particular as regards the impact of technical innovation and the new scientific approach to social issues.

The real beginning of the debate, however, has to be traced to a thesis put forward by Henri Pirenne in his *Histoire économique et sociale du Moyen Age*. According to Pirenne, the most important factor in the transformation of the rural classes during the twelfth and thirteenth centuries was not only the extraordinary population growth but, rather, the newly developing cities and the growth of commerce. It seems reasonable, then, to expect that a feudal organization quite fit for an epoch in which lack of outlets made it compulsory to rely exclusively on internal consumption would come to an end when permanent urban markets were established. From this point of view, the fact that feudal organization appears to fall into a state of decay as trade develops is highly significant. Some maintain, on the contrary, that the reasons for the crisis and final transformation of the feudal economy were mainly internal to feudalism. In this respect, Marx's explanation of the crisis of the feudal system appears as clear historical evidence of the structural contradiction between the existing type of social relations and the development of new productive forces, claiming a different relation between masters and workers.

It is difficult to exaggerate the importance of the Marxist position for the various schools of economic thought. However, it is surprising that such a position has been mostly accepted or rejected on philosophical, institutional and historical grounds, while the technical aspects and the underlying scientific issues have been generally blurred, if not completely overlooked. Why?

The answer must be sought in Marx himself. He suffers from an intrinsic contradiction, which he was never able to see, even less to solve. For Marx, as I have elsewhere observed, the basic connotation by which a society may be defined is not, it seems, its technology but its 'social structure', or, to put it more precisely, its 'class structure'. Although inconsistently expressed, depending on whether we look for it in his systematic works or in his occasional, polemical contributions, the idea of 'class' and 'class structure' is for Marx the fundamental concept. Were technological factors to play a determining and really crucial role in historical development, the first consequence would be the dropping of the dialectical approach, since technological development cannot be analysed by it or, more simply, does not permit that approach. Its character is cumulative and evolutionary, not dialectical. Its analysis rests on sociological and experimental research, and eludes a dialectical historical approach. The most intellectually daring attempt to unravel the Marxian contradiction in this respect has been made by Kostas Axelos who tries to interpret Marx as a 'thinker of technics' without fully realizing the contradiction between technics and dialectics.[4] In fact, Axelos's intention is clearly stated as aiming at an analysis of Marxian thought as a 'descriptive and dialectical interpretation' of technics as a human, voluntaristic initiative:

4. K. Axelos, *Marx penseur de la technique: l'aliénation de l'homme à la conquête du monde*, Paris, Editions de Minuit, 1961.

> Men were naturally prone to division of labour; in the future, they will be in a position to devote themselves *voluntarily* to social activities, going beyond the suffocating framework of the division of labour. . . . One should never lose sight of the very strict connection which binds division of labour to property and the different forms of division of labour to the different forms of property.

This attempt has merits. It is true that it seems to overlook the internal contradiction of Marxian theory. But, in a cultural situation in which Marx has been usually commented upon *sub specie philosophica*, it underlines the peculiar nature of Marxian analysis, which is neither deduced from Platonic ideas nor neo-Kantian categories but is rather outlined and set forth in its day-to-day unfolding through practice and the instruments of practice:

> Neither *politics* nor *religion* nor *art* nor *philosophy* are for him [Marx] constitutive forces of man's history (even alienated) of the ways in which man and the world are bound together. Practice alone is the source of truth and reality; it is practice that binds man to the world.

This is very well and convincingly put: practice with its instruments, that is, technology. More than a comment on Marx, it is an inspired repetition, which fails, however, to see the intrinsic contradiction of Marxian thought between technics and dialectics. We read, in fact, in the first volume of *Das Kapital*:

> Technology discloses man's mode of dealing with Nature, the process of production by which he sustains his life, and thereby also lays bare the mode of formation of his social relations, and the mental conceptions that flow from them. Even every history of religion that fails to take account of this material

> basis is uncritical. It is in reality much easier to discover by analysis the earthly core of the misty creations of religion than, conversely, it is to develop from the actual relations of life the corresponding celestialized forms of those relations. The latter method is the only materialistic and therefore the only scientific one. The weak points in the abstract materialism of natural science, a materialism that excludes history and its process, are at once evident from the abstract and ideological conceptions of its spokesmen whenever they venture beyond the bounds of their own speciality.

Far from opposing mechanistically structure and overstructure, Marx insists that one should take a hard look at the 'practical basis' of human existence. No doubt there is in him a large portion of 'dialectical impatience' of Hegelian origin, but, whenever Marx devotes his attention to specific historical phenomena, (from the civil war in France to the situation of the working class in England), his critical realism is impressive. In the first volume of *Das Kapital,* especially in the chapters dealing with the working day and with the mechanization of industry, his description of factory life is excellent and should be considered essential to any scientific sociology of work. For instance, with one sentence Marx disposes of the hard-dying legend of the steam engine as the mother of the Industrial Revolution, nor does he substitute for it the clock, as Lewis Mumford would have it. Marx sees with astonishing perceptiveness that the revolutionary machine is actually the machine tool. This is the machine that made a revolution in the form of steam engines necessary. As soon as man, instead of working himself with a tool, becomes merely the motive power of a machine tool, it is a mere accident that 'motive power takes the disguise of human muscle: it may equally well take the form of wind, water or steam'.

For Marx, practice is not an absolute point of departure, a pre-categorical postulate. It is, on the contrary, the specific life-situation. It is not, as in the Heidegger of the 'the question of technics', technology in general *sub specie aeternitatis*. It is the actual 'factory system'. More than that, although this is not only implicit in most thinkers dealing with technology as a global social phenomenon, the factory system is seen as a real system, that is, as a complex of rules and regulations, which means essentially that technology is both a practice and at the same time a power structure. Hence, it is possible to understand that, while the Industrial Revolution might not be a revolution in the proper sense after all, it points to a break with traditional routine and it forces a new allocation of power between different social agents. Naturally industrialization as a global social process involves states of mind as well as organizational, structural aspects of society. It would be idle to speculate on the relative priority of one set of aspects versus another one – for instance, psychological aspects versus objectives ones – it seems clear, however, that structural problems cannot be reduced to a personal headache.

A more recent debate on the Industrial Revolution and its impact is taking place. Brenner's argument, for example, is simple enough: if agricultural productivity is important for the transition to capitalism, then why did this transition first occur in England? Brenner's answer is still basically a Marxist one. In fact, it is his contention that in England the tendency was towards the formation of larger and larger farming units, the consolidation and the leasing of them to large tenant farmers who would hire rural wage labourers. According to Brenner, the emergence of this peculiarly English structure of class relations made possible the great increase in English agricultural productivity and this, in turn, sparked the overall economic development. The debate is still going on. Some argue that Brenner gave too much importance to agricultural productivity, especially to the difference in rural productivity between England and France; others contend that technological factors rather than class factors have to be isolated and regarded as the decisive ones while some commentators still insist that agricultural productivity could not be considered as a special and exclusive quality of large-scale capitalist farmers. All these views have some merit, but the fact remains that Brenner's argument, valid as it is for Europe, cannot explain the rise of capitalism among the English colonies on the eastern Atlantic seaboard of North America. In fact, while no large-scale capitalist could be found in eighteenth-century North America, it is undeniable that capitalism developed precisely in those regions, like New England, where small-scale family farming was the dominant form of economic activity.

Evidently, cultural traits as well as psychological and religious motivations together with structural features of the economic scene must be taken into consideration. In this respect, Weber's and Sombart's contributions still have some value. They have to do with a typical misunderstanding of the nature and the role of technics in economic and social development. We take the term 'technics' in its wide meaning to include scientific inventions and their large-scale application, and as such we can use it as a synonym for technology. Technical progress has been so far largely misunderstood. When, confronting the problem of the machine, Heidegger asks what the essence of a modern machine is, the answer he offers is far from satisfactory. He maintains that a machine, and in general technology, cannot be anything else but a new form of the eternal return of the identical. We have already remarked elsewhere that, expressed in these terms, the problem posed by technology cannot escape a decadentistic outcome, that is to say, one that would confuse important analytical observations with purely suggestive propositions. Needless to say, such propositions, oracular and suggestive as they are, can be neither verified nor falsified and of necessity lead to the elaboration of the myth of technics – a myth that stems from an equivocal experience, because it expresses an evaluation of technics as though technics were completely alien to us instead of being, as they are, man-made and historically determined. This is not to mention the fact that anti-machinism is usually the product and attitude of people unfamiliar with machines and who entertain an irrational fear of something they are simply ignorant of.

Marx, as we have seen above, offers a clear and factual description of technical processes and seems now and then to realize the revolutionary impact of machine technology but he fails to clarify his views about man's work when he states that such work essentially consists of some sort of 'metabolic process' which involves man and nature. What Marx misses has to do with the very nature of technology. For him, technology, more than a perfection without purpose, as it has been eloquently argued, has a purely instrumental nature. What both Marx and Marxists fail to see is that technology is not only a mediating tool in the production process. Technology in the present-day factory system has actually become a direct productive force, and science, on the other hand, from which technology derives in the form of applied scientific knowledge, is directly fulfilling a social role and affects both the nature and the allocation of power and resources. Science and knowledge are fast becoming decisive social factors. What one is in a position to control at the present time is only what one knows.

Strange as it may seem, perhaps the only major thinker of the past century who actually managed to conceive and elaborate the social role of science is the official founder of sociology, Auguste Comte. Far from being a lonely and mysterious enterprise as the popular imagination still sees it, science in Comte's view fulfils a basic social role and is a determining force in modern society. Again, such a role is not merely instrumental. It has to do with the basis on which modern industrial society legitimizes itself. This legitimation cannot be achieved by tradition, by metaphysical or by technological values, but is necessarily based on scientific knowledge. What defines modernity is precisely the new concept of science – no longer a solitary initiative or esoteric, mysterious knowledge for the happy few, but the social foundation of any society that is daring enough to embark upon a process of renewal, cutting off its old values and beliefs. For Comte, science is not only knowledge, to be conceived as the product of 'idle curiosity', it is knowledge in terms of social process, planning and initiative. How is this possible? The main lines of Comte's reasoning are simple enough. His teacher, Saint-Simon, had stated that 'mankind is not fit to live in ruins; it needs certainty and stability'. After the French Revolution, mankind – *l'humanité* – had to rebuild its habitat; society needed total reconstruction. New values had to be discovered and formulated to function as the basic presuppositions for the various social groups. Whence could these values come? Tradition was over. Old-fashioned theology and speculative philosophy had little, if anything, to offer. Individual wills, on the other hand, were too unpredictable and drastically differentiated to provide a common basis for the new community. Social consensus had to be founded on a totally new basis. Here Comte's contribution emerges as a stroke of genius. Individual will is anarchistic in nature and therefore not to be trusted. But science, on the contrary, has an intersubjective binding value. Science does not depend on individual principles of preference. Science is a public procedure. It is convincing with the pure force of reason. It can act as the new basis for social consensus.

With science as the basis for social cohesion and as the binding force of the new community, the historical process, which had begun with the early phases of the Industrial Revolution, seems to have reached its logical outcome. It has been aptly observed that historians used to call it the 'Industrial Revolution'; some social scientists label it 'modernization'; the economic anthropologist Karl Polanyi termed it the 'Great Transformation'; Marxists called it 'the transition from feudalism to capitalism'. But by whatever name it has been known, we have been fascinated with the emergence in the late eighteenth century of the great industrial or capitalistic market economies of the Western world. Now with the collapse of communism in the former Soviet Union and Eastern Europe this fascination has taken a new relevance. Can what happened in the West perhaps be useful today to the new emerging countries and developing economies all over the world?

What has to be understood is the new social function of science and of technology as applied scientific knowledge. We still talk and think in eighteenth-century philosophical, economic and political terms – from capitalism to socialism – and do not realize that the new social realities can no longer be accommodated within these categories. The ideological conceptual framework of the eighteenth century is obsolete. It must be realized that science, scientific knowledge and technology as applied science are the new faces of social and political power.

The role of information in present-day developing societies is crucial. Information has become the essential precondition of human development. During the eighteenth and nineteenth centuries, exploitation of man by man was direct and more or less precisely quantifiable. It had to do with muscular energy, working hours, wage levels, factory disciplines. The scenario today is vastly different. The description in Volume 1 of *Das Kapital* is no longer tenable. The new variables are isolation, segregation, solitude, neglect, exclusion. To be exploited in the modern world means to be on the fringes, to be cut off. Our everyday experience and common sense might seem to contradict this statement. The mass media continue to spread all over the world. Decisions, at all levels of organized association, are more numerous, more clamorous, more widely known. But access to a centre where these decisions are taken is more difficult. The centre is remote. Its arbitrary nature is consecrated by mystery. It is even hard to determine where exactly this centre is located and who forms part of it. The message of power risks becoming Kafka's 'message of the emperor'. Let us set aside the philosophical and sociological theories by which the various power élites defend themselves. Then a curious, and in many ways unexpected, fact must be pointed out. Power today exploits and oppresses not by the use of direct action, but simply by ignoring, by failing to intervene, by refusing to take action, by taking refuge behind complex and perfectionist procedures through which legal formalism and political paralysis come to each other's aid. The most serious sins of power today are sins of omission. The genuine reactionary today is not the man who has a gallows erected or applauds the censor, but the man who prevents action, who preaches resignation at all costs and urges men to put their trust in a spontaneous, automatic evolution and practises a 'benign neglect'.

In this connection, social information has at least three basic functions: (a) it prevents popular initiative from dying away as the result of bureaucratic complications and static self-enclosed institutions; (b) it guarantees, up to a point, the effectiveness of the pressure from below as a major democratic instrument capable of ensuring people's participation in the process of social transformation; (c) it tends to maintain a degree of contact between political decisions and citizens' aspirations, protecting them against both authoritarian rashness and paternalistic tendencies.

These three functions can be fulfilled not by social information but by a type of information based on scientific analysis. It is here that we can fully understand the new role of science as a socially meaningful enterprise, as well as the great contribution of Auguste Comte and his motto: '*Savoir pour prévoir; prévoir pour agir.*' Science and technology are essential for the perception and interpretation of present-day society. This society cannot be merely described as an information society. We need to be more precise. It is a society based on scientific information, that is, information that is intersubjectively binding. A society today can be capitalistic or socialistic or co-operativistic or based on a mixed economy from an ideological and institutional point of view. However, the underlying fundamental criterion for its definition goes beyond these external features. It points essentially to the way and to the degree in which science and technology tend to shape its structure and its basic developmental orientation. In this perspective, the importance of science as such, of science in its broadest sense, including natural and social, or moral, sciences, becomes clear and the famous polemic about the 'two cultures' loses much of its value.

The pamphlet by C. P. Snow is well known. My feeling is that it has been grossly overrated. To understand the full impact of modern science we should perhaps take into account the concept and practice of science in Classical Antiquity. After the successful defence of his home town, Syracuse, Archimedes would burn his notebooks with the practical lessons from his experiments. The classical scientist did not have much appreciation for practical, useful application. Slaves were available in great numbers. One did not have to have resort to machines as labour-saving devices. Moreover, among scholars the fear of menial contamination was widespread and had an arresting effect. It has been properly remarked that Greek science did not arrive at a great technological development because it did not want to reach one. In Classical Antiquity and the Middle Ages, the scientist kept quiet and by keeping quiet science made the state, the political and religious powers, afraid. It won respect, if not money. Scientists were persecuted, tortured or killed, but in general they were able to preserve their freedom. It is a fact, as we have seen in the previous chapters, that men of power can live, fight, and actually be powerful, only so far as the means offered by scientific culture are available.

As soon as science becomes something different and more than pure speculation, and the man of science is ready to offer his knowledge for social use, especially in the growing field of industrial production, the situation changes. The scientist is no longer independent. He has traded his autonomy for a position close to power and for the material means to conduct his research. He is no longer a loner; he works more and more in a group, is part of an organization. Science responds to a compelling question of the society being transformed. Traditional criteria of judgement are undermined by widening communications and the loss of legitimacy of the ruling powers. Once tradition as a source of legitimacy has fallen, to whom can one turn? Where can one identify a new instrument of self-listening and self-direction for societies which were so adventurous or imprudent as to take the paths of modernization on a vast scale? Where can one find a new basis of social consensus?

Science is the almost obvious answer. It used to be said that the cultivation of science was an act of homage to God, in so far as it permitted the study of nature, God's own creation. Newton considered God as the Supreme Architect, the 'pantocrator' and guarantor of the cosmic order. However, modern science seems to possess a unique hubris, unwilling to limit itself to reflecting on nature. It proposes to imitate nature but at the same time invade it, exploit it, reinvent it to the point of violating it. Thus, the humility of the scientist as regards nature is merely apparent; it is a tactical move to disarm the adversary and take him by surprise. Bacon had said: '*Natura non nisi parendo vincitur.*' But Bacon's 'obedience to nature' was only the humility that precedes and prepares the triumph over and against it.

The 'sanctification' of science found an unexpected reinforcement in the Puritan ethic, which was based on utilitarianism and on the idea that all knowledge should be evaluated, contrary to the wisdom of the classics, on the basis of its utility. In fact, it was thought that everything that helps to make the life of man happier and less hard is good in the eyes of God. Naturally – and on this point the utilitarian and Victorian moralities were logical to the point of cruelty – this is to be understood as applying only for all those who deserve it or who pay for their welfare and their 'good standing in the community' through methodical daily labour, diligence in undertaking their professional duties, and total dedication to their task. Thus, the whole of social life, both individual and collective, is permeated by a scientific spirit and rational calculation. Instrumental reason has at present gained the upper hand in the developed societies and the newly developing countries alike. It could be said that Auguste Comte is finally enjoying some sort of posthumous revenge.

Science and scientists often serve the existing powers. And, in fact, they do serve. But in so doing, they change the nature of power and those who wield it. The very source of power and the ways in which power is being exercised are more and more deeply affected by science and technology. Even traditional property rights – such as the private property of invested capital – are undermined by science to the point of fast becoming obsolete as source of power. In large enterprises at present there is a clear divergence between the power stemming from legal property rights – the shareholders – and the power from effective everyday control of operations – the managers. Moreover, capitalism can no longer be defined simply by taking into account its rational book-keeping methods, as Max Weber maintained. Capitalism can be defined even less in Werner Sombart's terms as a 'predatory enterprise'.

Present-day entrepreneurs have little in common with the romantic innovator whom Joseph A. Schumpeter used to celebrate a few decades ago. In his book *A Theory of Economic Development*, Schumpeter seems closer to Friedrich Nietzsche and Max Weber than to Adam Smith or David Ricardo. In the first place, he sees in the entrepreneur a dream coupled with the strong determination to find a private kingdom. In the second place, Schumpeter elaborates on the idea of conquest; in his opinion, there must be in the entrepreneur the impulse to fight, to beat his competitors and to show everybody that he is the best in his field; finally, just like a poet or an artist, the true entrepreneur must be sensitive to the 'joy of creation'. In this view, the entrepreneur cannot wait for a quiet automatic development of the situation. Contrary to what the classical economists used to think, he cannot expect the market to force him to develop, to take a dynamic stand. He must himself create the market. He must anticipate developments, needs and values that are not yet there. In this sense, far from just being a good administrator, the true businessman in Schumpeter's opinion is a charismatic leader, an inventor more than a routine organizer. This romantic portrait has not much to do with business activity as it is conducted at present. Innovation is more than ever necessary. But innovation is no longer the irrational gift of the gods nor the product of personal charisma. It cannot be divorced from scientific planning. It is a function of rational calculation. It rests on empirically tested knowledge. Moreover, only exceptionally can it be found and organized by a single extraordinary individual. Innovation as well as scientific research have largely become a group activity; they need continuity, a multidisciplinary approach, constant cash flow in an integrated financial plan, and implementation and evaluation of specific results.

In this sense, the Industrial Revolution, and capitalism stemming from it, determine the emergence of a new society. This society is new not only because it provides new products and new ways of life. It is new because it is based on a new concept of power and because it is endowed with science replacing traditional values and transcendent religious dogmas as a basis for moral obligations.

Today's advanced societies are therefore essentially characterized by the all-pervading power of scientific knowledge. From a historical point of view this is a new phenomenon. Its consequences on society as a whole and on science itself are interesting but not easily detectable or fully understandable. Once science has become a fully organized group activity, it tends almost inevitably towards bureaucratization, with an obvious stultifying effect on individual inventiveness. Everybody knows that any scientific research project has at least two phases: first, a 'context of discovery' in which the researcher or the research group relies essentially on his or its 'poetical imagination', as it were; then, in a second phase, there is a predominant development of the 'context of validation', in which both the general orienting hypotheses and the specific working hypotheses are duly tested on an empirical basis. Taking this peculiar structure of scientific research into serious consideration, it becomes impossible or simply a matter of idle speculation to write and talk about 'the two cultures and the scientific revolution'.

Within political parties, pressure groups, trade unions, and so on, research is often called upon to supply data to give policy-makers a good chance of cloaking decisions that usually have already been taken with scientific jargon. This amounts usually to an exercise in petty rationalization. Science is being used as a polemical weapon against rivals and competitors.

It is a different story when science works for the government or for an international agency, sponsored by various governments. Then a strange paradox seems to take place. Science is undoubtedly a tool of government, but at the same time government and its power are somehow coerced into its service. Between science and power a curious master-servant relationship develops. One has to go back to Hegel: the servant serves his master but by serving his master the servant little by little empties him, makes him devoid of any original initiative like an absentee owner, reduces him to a pure façade and finally replaces him. Government grants money to science to achieve certain goals, to conduct research with no knowledge of technical details and procedures and therefore lacking effective control. Thus, science in the end dictates to government what to do, how, and how fast. No control is possible without actual knowledge. In the end, policy-makers, despite all their rhetoric, are powerless when squarely faced by scientists.

There can be no doubt about the increased social significance of science in modern society. To think that the problem of interpretation of this society is solved by calling it 'science' or 'knowledge' or 'knowledgeable society', begs the question. The object to be explained becomes the key to explanation. The great problem of technically advanced present-day societies remains unsolved.

Knowledge, all types of knowledge including strict scientific knowledge, are not *per se* self-revealing. Like technology, knowledge cannot be regarded as a *deus ex machina* to be called upon to solve the social problems that it has very often itself created. Like technology, scientific knowledge is essentially perfection without purpose. Thus, the basic questions to ask are: Knowledge from whom? For what?

We are thrown back to the underlying problem of power and its substantive legitimacy and its value orientation. Science can clarify the issue, calculate the relative costs of the various alternative courses of action, show the best ways of implementation. But it cannot take away from us the human responsibility of choice and decision.

But going back to the origin of the Industrial Revolution, a good example of the historical impact of philosophy and science on power positions can be seen in the life of Diderot and Voltaire. Most of the writers of the *Encyclopédie*, especially D'Alembert, were asked for advice by persons in power, both political and social. These included Christina of Sweden and Catherine the Great of Russia, The position of the savant, however, was generally one of subservience. He was supposed to offer opinions and even plans for action but he could not expect to see them implemented and, if so, even less was he given permission to control the process of implementation.

With respect to ancient philosophers and scientists who would refrain from making public their findings, the writers of the *Encyclopédie* were quite conscious of the social impact of their thinking. Learned publications, in this respect, had to be written in a style clear enough to be understood by everybody and to have an immediate impact. One book could actually help another publication. Voltaire's famous *Dictionnaire philosophique* (1764) seems to have the obvious aim of helping and accompanying, as it were, the major *Encyclopédie* of Diderot and D'Alembert. In fact, Diderot was quite conscious of the difficulty that most contemporaries would encounter in understanding the full implications of his intellectual position. In his Foreword, Diderot writes that 'even the most enlightened person will find in here ideas that he does not know and facts that he is ignorant of'.

Although less radical than Diderot, Voltaire had a more direct influence on the growing 'ideology' of the bourgeoisie. When about to die in Paris in 1778, he could whisper to Benjamin Franklin, who had come for a visit: 'God and Freedom', as a kind of last wish.

Friedrich Nietzsche once remarked that among all the cold monsters the state is the coldest one. But science, at the present time, needs precisely that monster. Science has become, especially after the Second World War, a huge enterprise. It is no longer an individual, lonely pursuit. It has become 'big science', and big science, in order to function, needs a permanent organizational structure. This means that science needs big money and continuity. It cannot depend on the intermittent help from occasional patrons. There is probably not a single scientific project today that can be developed without government money. On the other hand, no major government decision can be taken and implemented without preliminary scientific research and constant scientific monitoring. This is the most important reason why science has become the new source of power legitimation while, at the same time, state structure and government power depend more and more in their day-to-day operations on scientific research.

There is practically no field of government activity or of private industrial development in which scientists are not making a mark. Undoubtedly, as it has been observed, there is often an intricate, complex knotting of public and scientific interests. A revealing, but not well-known, example can be found in E. Wenck's *The Politics of the Ocean.*[5] The author, a geographer, had worked at a high level in the Federal Government. In his book, Wenck describes in detail his efforts to enlist

5. E. Wenck, *The Politics of the Ocean,* Seattle, University of Washington Press, 1972.

scientists from a staggering array of specialities, including the White House itself, other governmental agencies and officials, Congressmen, and various public groups in support of a programme of oceanic investigation. Without the complex web of negotiation and persuasion that he was able to weave, this programme would never have been brought into existence. The same or analogous examples we can derive from Bruno Latour's *The Pasteurization of French Society*,[6] in which it is possible to learn how Pasteur manipulated opportunities within various social and medical milieux in order to advance his own scientific work and at the same time his career. Although the order of magnitude is at present a formidable one, the phenomenon is nothing new and goes back to the beginning of modern science in the eighteenth century.

6. Bruno Latour, *The Pasteurization of French Society,* Cambridge, Mass., Harvard University Press, 1988.

Science and technology since the end of the Second World War

Alexander King

The years since the end of the Second World War have seen the consolidation of today's technology-based materialistic society. As Dennis Gabor, the Nobel laureate and discoverer of holography, has remarked 'our present civilization is based materially on an extraordinarily successful technology and, spiritually, on practically nothing'. Not only has technology shaped our society, but discoveries in the scientific research laboratories have shaped technology.

During this period there have been many changes in the aims and emphasis given to research-and-development (R&D) and in the attitude of the general public towards science, which has veered between veneration of its mysteries and distrust and fear of its powers for evil. In the post-war era we can distinguish four rather well-defined phases in the relationship between technology and socio-economic climate.

The first phase lasted until the early 1960s. It was a time of hope for a peaceful and constructive future. The applications of science had played a visible and decisive role in the war, and the general public, especially in the victorious countries, believed that it could be equally effective in the new peaceful world they believed was emerging. The politicians shared this view and there was a massive increase in the resources made available to R&D in many countries. This was, indeed a period of euphoria for science and the scientists. Nevertheless, general strategic considerations and the emergence of the Cold War dictated that a large proportion of the research and development effort should go to the military. Physics and the technologies built upon it held centre stage. The public tended to view the great physicists as Olympian and detached. Realization that Oppenheimer had a conscience came as a surprise.

With the completion of post-war reconstruction and the consolidation of the peacetime economies came the second phase which lasted until 1967. It was characterized by unprecedented economic growth in the Western countries and Japan. The importance of R&D in the generation of this growth was tardily recognized and much academic effort went into the exploration of the science-technology-production relationship. This phase was essentially one of attempting a rational approach to the applications of science and the allocation of resources to it. Strategic considerations and the Cold War sustained defence R&D at a high level, but even here much attention was paid to their spin-off value to economic and social development. For the first time R&D statistics were systematically collected both with regard to government expenditure and that of industry. This enabled international comparisons to be made of the level of R&D as a proportion of gross national product (GNP), indicating considerable disparities between the scientific efforts of different countries. The dominant position of the United States in this area led to concern in the European countries that the 'technological gap' might seriously threaten their competitive position. For industry, similarly, inter-firm comparisons of R&D expenditure stimulated the efforts of individual enterprises.

During this period, understanding of the role of science and technology in the economy was at a very naïve level, but it soon became accepted that investment in R&D was necessary both at the national level and that of the individual firm. This was an era in which attitudes towards science and technology were positive and hopeful. After all, it was vaguely sensed that they were partly responsible for the new prosperity from which so many were benefiting. The world was in trouble, certainly; the threat of nuclear war loomed large and conflicts were erupting in many Third World countries, where poverty and hunger held sway. But this might well be due to too little or the wrong kind of science, or to ignorance of how to apply it.

The third phase was one of disillusion with science and technology. In the minds of many, scientific research had become associated with war and the preparation for war. The Cold War could be seen as a competition between the scientists and engineers of the two sides to create even more complex and sophisticated weapons of destruction, the nature and potentialities of which were well beyond the understanding of their own political masters. Likewise signs, too important to ignore, began to appear of environmental and social deterioration resulting from the large-scale application of technology. Environmental protest became widespread and governments and industrial companies finally had to take it seriously. Economic growth was seen by many as a mixed blessing; governments had, in fact, neglected to consider its social and human consequences. The result was a spread of alienation, a sense that the material rewards were shallow and that quality was slipping out of life.

All this led to a general questioning of the aims, orientation and even the methods of science. Scientists began to lose their credibility with the public; they were seen as instruments of military and economic domination and unconcerned with the pressing social problems of the times. Science was seen as élitist in its organization and mores; its vaunted objectivity was met by derision in some quarters. Much of this disenchantment turned also against high-technology industry; large corporations and especially the transnationals were regarded with great suspicion. For the first time for many years, the steady, and at times spectacular, growth of R&D began to fade.

The fourth phase started with the oil crisis of the early 1970s and has continued to the present. It has been a period of slow economic growth and many uncertainties. The oil crises demonstrated how vulnerable the industrialized countries were to potential interruptions of supply of energy and raw materials on which their economies depended, caused by political or military events beyond their control and difficult to foresee. The new wealth of the oil-producing countries led to extensive recycling of investments. Competition between companies and countries hardened considerably in markets which were sluggish and in some cases near to saturation. There were innumerable mergers and take-overs among corporations, mainly aimed at increasing or at least maintaining a market share. In this process innovative potential was at times disregarded. Heavy industry declined considerably in the older industrialized countries, while the incursion of the Japanese into the already inflated automobile and electronic industries of Europe and the United States caused much damage but evoked little innovative response.

New technology during this period arose from the application of scientific discoveries, chiefly in two fields – solid-state physics and molecular biology. The development of the micro-electronics industries has been rapid, highly diversified and very dramatic. Computers and devices based on miniaturized integrated circuits are all around us; automation and robotization are transforming industry and the services with massive impact on the economy and also on lifestyles. The Industrial Revolution made it possible to replace the puny energy of men and animals by steam power leading to the transformation of society. The development of micro-electronics means that a brain and memory can be provided for any piece of equipment devised by man, at a modest cost; this is the entry point to the post-industrial society. Micro-electronics links up with a complex of other advanced technologies which include satellites, liquid-crystal and glass-fibre technologies, and holography.

The economic and social consequences of the micro-electronics revolution are considerable. It is leading humanity into a post-industrial, information or service economy, with striking changes in the nature of work and patterns of employment. It would appear that a large proportion of the present levels of unemployment in the industrialized countries is due to the automation of industries and services, and not merely to the cyclical effects accompanying the recent recession. Yet, as a means of combating unemployment, governments everywhere, in attempting to increase competitivity by raising productivity, are essentially encouraging further technological development and hence job losses. Such an approach leads to the creation of more wealth, but destroys manual work – so called jobless growth.

The biological counterpart of this development has been slower in its applications, but is even more fundamental. The discovery of the structure of DNA, the unravelling of the genetic code and methods of inducing mutations, together with advances in cell and membrane biology provide new possibilities for the understanding of life and indeed of modifying it. Genetic engineering has already made considerable contributions to medicine and agriculture. It has immense possibilities, but raises difficult ethical problems.

The period has also been one of great environmental uncertainty, exacerbated by a rapid rise in the global human population, unique in history. The discovery of worldwide pollution problems such as the depletion of the ozone layer, the spread of toxic materials and the increase in the so-called 'greenhouse effect', which might lead to global warming, have called for a similarly global response; they cannot be tackled by individual countries in isolation. Their containment calls for new forms of international co-operation in research, monitoring and action. These environmental problems are vivid reminders of how scientific exploration or discovery alters the human perception of its predicament, with heavy impact on the behaviour of industry and also on the policy of governments, for example in energy and transportation.

Finally, this phase has seen the collapse of the Marxist societies and economies, the ending of the Cold War, and the greatly reduced threat of nuclear holocaust. The ending of ideological confrontation was, of course, seen as a great step towards global harmony but has already produced its own set of acute problems. The end of the arms race has given rise to a considerable reduction in defence expenditure, but it is not yet clear how far defence R&D has been affected. Since success and even survival in future conflicts that threaten war will depend critically on superiority in sophisticated weaponry and delivery devices, the development of military technology is bound to remain a preoccupation of all countries that can afford it.

The scientific legacy of the Second World War

The use of science in the First World War was relatively primitive, but between the wars considerable developments had taken place. The two main science-based industries, chemical and electrical, had flourished and given rise to corporate giants such as I. G. Farben and Siemens in Germany and Dupont and General Electric in the United States. Scientific research was also stronger, undertaken mainly in the universities and the great laboratories of some governments. Some of the research schools of Germany were especially famous. It appeared, therefore, that both sides in the war had roughly comparable scientific efforts. In effect the conflict was decided by the superiority in the use of science by the Western allies, probably because Germany, convinced that the war would be short, had dispersed many of its scientists at the outset.

The innovative technology thrown up in the war effort was impressive – the jet engine, the early computers, rockets, radar and a great range of electronic devices, proximity fuses, penicillin and DDT, among many others. Above all the other technological successes towered the achievement of the atomic bomb. Its importance in the abrupt termination of the war with Japan was minor in comparison with its influence on postwar geopolitics.

The Manhattan Project which produced the bomb was a remarkable example of co-operation between scientists and engineers, amply funded but working against time to achieve a major national objective. Starting virtually from theory, the project succeeded in surmounting all the problems of fundamental and applied research as well as technological development sufficiently quickly to achieve a bomb before the war was over. It was the model for the only other mammoth technological project mounted by a government, the placing of a man on the moon three decades later.

Wartime technological development was based on effective co-operation between science, government and industry. For the Western allies, it involved a complete mobilization of the scientific community. It was surprising how easily scientists normally engaged in academic research were able to turn to applying their skills to the development of the devices of war. Relations between the newly recruited scientists and the quite small and traditional technical services of the military were not too easy at the beginning. But full collaboration was soon established. Well-known scientists also appeared as advisers to ministers, chiefs-of-staff and other decision-makers, again a pointer to the future.

The place of industry in this situation was, of course, crucial, not only in production, but in designing chemical pilot plants and engineering prototypes and, by contract, undertaking responsibility for particular development projects. In many instances, too, corporate scientists and engineers were outstanding innovators. The skills developed by industrial firms proved valuable after the war as did the spin-off from defence development to innovations in the civil economy.

A particularly interesting war development was the appearance of operational (United Kingdom) or operations (United States) research (OR). This was essentially the use of simple statistics or often just scientific commonsense to real operations. OR groups were attached to particular commands. Its most dramatic success was in the Battle of Britain, when OR, working in association with microwave radar which the Germans did not possess, enabled the Royal Air Force to defeat a much larger air force by foreseeing situations favourable to combat and ignoring unfavourable conditions. It is interesting that the successful OR scientists came from widely different disciplines, such as physics and plant genetics. What was important was a sound feeling for the scientific method and the establishment of a relationship of mutual confidence between the OR scientist and the military decision-maker. This technique had a considerable vogue in industry during the immediate post-war years.

It may be useful here to insert a digression concerning the qualities desirable in science advisers to political or military leaders. The adviser is not necessarily a star scientist, and he must not pretend to know all the answers. Indeed, although there will be exceptions, he is unlikely to be a world name in science, but rather a person of broad scientific knowledge and commonsense in contact with a network of colleagues from many scientific disciplines whom he can consult on specific points and endowed with a personality capable of creating a relationship of understanding with the leader he serves. At a meeting held in the 1970s of the science advisers of the various American presidents, it became clear that in only one case, that of Jerry Wiesner and President Kennedy, had such a relationship of mutual confidence existed to make advice systematically effective.

An interesting case is that of Sir Henry Tizard who, as chairman of the British Aeronautics Research Council before the war, was responsible for the rapid development of microwave radar and advanced fighter aircraft.

When Churchill came to power, he brought in his own personal science adviser, Professor Lindemann, later Lord Cherwell ('the Prof') and Tizard was forced to resign. After the war, the novelist C. P. Snow in his Gwatkin lectures at Harvard set out to show that if Churchill had come to power six months earlier and provoked Tizard's resignation then, the Germans would have won the Battle of Britain and history would have been very different. This indicates the possible dangers of critical decisions being taken on the advice of a single scientist.

The years of reconstruction

Science had been outstandingly successful in winning the war; it was essential that it should be equally effective in the peacetime economy and society. All over the world countries began to reassess their scientific resources with a view to reforming or creating a scientific infrastructure appropriate to the new situation. In the United States, where industry was so successful and so quick in exploiting new technologies, it was realized that many of the basic developments that had helped to win the war had originated from scientific research in Europe. It appeared necessary, therefore, to strengthen the basic research systems. President Roosevelt wrote to Vannevar Bush, the head of the exceptionally successful, but temporary wartime organization, the OSRD, asking him to put forward suggestions for the postwar equivalent. This resulted in the historic Bush Report, *Science the Endless Frontier*, which eventually gave rise to the National Science Foundation.

In the United Kingdom, the situation was the contrary. Many important discoveries made there seemed to be exploited abroad, with little economic advantage to the inventors. It was estimated that, while in the

United States, there were some 2.5 applied scientists and engineers for every basic researcher, in the United Kingdom, the ratio was only 1.2 to 1. In response to this, an Advisory Council on Scientific Policy (ACSP) was set up with, among other objectives, the task of studying how to achieve a better link between the scientific system and those of production and social development. There was much interest in the United Kindom in the cultural dichotomy which may have been at the root of the country's relative failure to make use of its scientific achievements, the co-existence of two élites, one of humanistic and the other of scientific education, who had almost ceased to communicate. It is interesting to note that a reprint of C. P. Snow's book, *The Two Cultures*, has just been published in London. Can it be that the situation still exists?

The mystique of science, inherited from its wartime successes, generated a rapidly mounting flow of resources in the industrialized countries. Initially, science was to some extent a *chasse gardée* to which parliaments voted resources with relatively little questioning, on the simple assumption that research is a good thing and that more research will solve more problems. Decisions were taken without much serious conflict in the easy circumstances of a continuous expansion, so rapid and sustained that it came to be regarded as an inevitable phenomenon, virtually free from political choice. The initial thrust of science policy in the OECD countries during this period of euphoria was essentially that of raising the level of R&D expenditure of the various countries, measured as a fraction of their GNP, to the level of that of the United States. In this process both governments and scientists regarded R&D as a valuable national investment in its own right, virtually without reference to the content, provided it was sufficiently creative and sophisticated, and with only superficial consideration of its relevance to specific national objectives. In this connection, it is interesting to recall that in the earlier of the periodic conferences of scientists and parliamentarians organized by the Council of Europe, the legislators wanted to be told about, and if possible visit, some dramatic examples of big science such as radio-telescopes and high-energy particle accelerators. They were captivated by the mystique of science, did not yet see the link between science and the economy and raised few questions about the economic yield of their investments in research. This was soon to change.

The British Advisory Council on Scientific Policy, created in 1946 and the first in the world of its kind, was taken seriously by the Government. Ministers sought its advice. For example, the post-war Chancellor of the Exchequer, facing problems of reconstruction, was hindered by the scarcity of hard currency. He was tempted to decrease food imports, which would have entailed the reduction of rations. He asked the ACSP if such a measure would have negative results by reducing the productivity of industrial workers. The council assembled a panel of experts in nutrition, ergonomics, etc., to tackle the question. They reported that a general reduction of the already low calorie input of the population would indeed have negative consequences, but that damage to production could be avoided by a differential rationing system in which manual workers were allowed more food than sedentary workers, housewives, etc. The Chancellor concluded that such a scheme would be impractical and politically dangerous, so the rations remained unchanged.

The Lord President of the Council, as minister responsible for science, posed a more general question. How might science best contribute to the economy and society in this period of reconstruction and capital shortage? A group of industrialists, trade-unionists and economists was formed to consider the matter. They were all convinced that research was essential in the creation of wealth, but as it was the nature of the science/technology system, many years would have to pass before new expenditure on R&D produced a harvest to be gathered. In the meantime, while reconstruction was far from complete, great efforts should be made in applying the scientific method to increase the efficiency of industry, agriculture and the service sector, particularly by raising levels of manpower productivity. Stress was given to the importance of the human factor by attempting to convince the workforce that it had a big stake in improving productivity and would share its benefits. Emphasis was also given to the need for upgrading management and for creating facilities for management education, which was at that time not yet accepted by academia. This triggered off the productivity movement. International and interfirm comparisons of output of production per man-hour or man-year varied greatly, exposing reasons for the differences and suggesting possibilities of improvement.

The dominant factor in the rebuilding of Europe was the generous and imaginative Marshall Plan. Through its agency in Paris, the Organisation for European Economic Co-operation (OEEC), the European members established a Committee for Productivity and Applied Research which generalized the British initiative in the creation of national productivity centres in each of the member countries. These operated on a tripartite basis between governments, management and labour. They proved to be effective at the grass-roots, particularly in the transfer of industrial techniques from the United States to Europe and in the establishment of management education centres in universities.

Two cases: Japan and the Soviet Union

So far we have discussed the post-war situation in Europe and North America. Things were quite different in Japan, a nation stunned by defeat, shocked by the atom bombing of Hiroshima and Nagasaki, and occupied by a foreign power. It had to regain confidence in itself and rebuild a society and an economy with new values and new objectives. In this it had great assistance from the United States and relief from the burden of military expenditure, but it was the national characteristics of solidarity, consensus-seeking, paternalism, will to succeed and an exceptional capacity for hard work which combined together to create the Japanese economic miracle. The late Saburo Okito told me once how, on the very evening of the armistice, he was one of a group of men who met in the office of the senior professor of economics of Tokyo University and drew up a strategy of development. The first step was the rebuilding of the education system including the creation of a large number of new universities with important faculties of natural science and engineering. The new research schools would not be in a position to supply the ingredients for indigenous innovation for many years, so industry first had to rely on techniques from abroad. These were selected with remarkable care. Many groups of experts toured Europe and North America, holding discussions with government agencies, industrial firms and individual scientists and engineers. They gained an extraordinary and detailed knowledge not only of existing technologies, but of those still on the horizon. On the basis of this, processes and products where chosen, rights acquired and production started in Japan. This was no mere imitation. In many cases products covered by foreign patents were redesigned to incorporate Japanese concepts of quality, accuracy and artistic design.

The construction of modern Japanese industry was undertaken through a remarkable co-operation between the leaders of industry and banking, with the Japanese ministry, MITI.

While reconstruction involved a wide range of industries, high priority was given to the electronics sector. Here again the Japanese displayed great foresight. Their long-term plans for investment in development and production were ridiculed in the West as unrealistic and hopelessly inflated, but time proved their validity. Electronics reaped rich rewards for Japan.

The situation in the Soviet Union was entirely different. Victorious in the war, but suffering from immense destruction, the USSR found itself in the aftermath of war with a mighty army and an extremely poor internal infrastructure. With its ideological zeal and its military power it dragged into its orbit the other countries of Eastern Europe, and imposed on them structures and policies similar to its own. The polarization between the communist system and that of the free-market nations inevitably led to confrontation. While the United States and its allies had sufficient strength to develop a sophisticated military technology to maintain a lead in the arms race and, at the same time to develop their vigorous civil economies, this was impossible for the USSR. Absolute priority was therefore given to military needs – guns before butter. Soviet science and technology was of a high standard. With the help of information of uncertain value provided by spies and Western traitors, the Soviet Union made it possible to produce both the initial atomic and the later hydrogen bombs in a remarkably short time. Furthermore, the high quality of the Soviet scientists and engineers was able to compete with the Americans in the development of the most sophisticated military devices including intercontinental ballistic missiles. The same ingenuity and skill were manifested in space technology. The successful launching of satellites by the Soviets came as a shock to the West and provided the necessary impetus to provide the resources for the American space programme. The civilian front, however, remained static and lacking in innovation. It is difficult to recall any novel product that the Soviet Union was able to sell on the world market. Thus economic growth remained low, as did standards of living. It is paradoxical that a country could exist in which military and space technologies reached unbelievable levels of sophistication, while people in its vast hinterland suffered a Third World type of existence. The organization of science in the USSR differed greatly from that of the Western countries. Research was undertaken mainly in institutes of the Academy of Sciences, rather than the academic environment. Members of the Academy constituted an important element of the élite and had special privileges, but many constraints.

Science and the economy

Until recently, understanding of the linkage between science and the economy has been at a very simplistic level. It has long been realized, of course that fundamental discoveries of science have given rise to great, new, research-based industries. Modern power engineering would be unthinkable without an understanding of the laws of thermodynamics; the electrical and electronic industries were not even conceptually foreseen before the work of Faraday and his successors; the pharmaceutical and other branches of the chemical industry derive from the molecular architecture made possible by the developments of organic chemistry.

Nevertheless, economists have until recently paid little attention to the role of science and technology in economic and industrial growth. Keynes, for example, refers only marginally to technology. It was assumed that technology and the research on which it is based is stimulated and brought into being by the interaction of economic forces, as one agent of Adam Smith's invisible hand.

In the early 1960s, serious investigations by T. Schulz, Sven Nilson and others on the factors determining economic growth opened the way for a better understanding of the place of technology. Denison (1962) in a study of the growth of the American economy since the beginning of the century, concluded that only some 40 per cent of the increase in the gross national product can be explained by additions to the traditional inputs of capital and labour. The greater part is ascribed to the 'residual factor', presumably consisting of a complex of elements among which education, science, technology, entrepreneurship and management skills have an important part. Essentially this and other studies indicate that it is the quality of labour and the quality of capital utilization that have enabled GNP to grow so rapidly. Manpower quality at all levels from unskilled worker to top management is raised by education and training; capital utilization is raised by technological innovation. While this study was hotly contested, its broad lines are generally accepted. Both education and science began to be regarded as national investments and as such were able to justify increased resources. Science is indeed seen as an autonomous force in the economy, indeed its main source of power.

The league-table rivalry generated by internationally comparable statistics of research-and-development raised questions in the late 1960s of what significance the R&D disparity between the United States and the countries of Western Europe had for the future economic performance of the countries concerned. The publication of J. J. Servan Schreiber's *The American Challenge* made this a public issue. Studies of the so-called technological gap indicated that there was, in fact, no direct correlation between a country's R&D effort and its economic growth or trade performance. It appeared that the diffusion of technology across frontiers, mainly in the form of the sale of patents and know-how, or comprehensive arrangements between industrial firms, had become so rapid as to compensate for inadequacies in the domestic research effort or indigenous innovation. The example of Japan, with its highly successful industrial development based on imported technologies, is a dramatic case in point. Nevertheless, effective transfer of technology does demand a level of educational quality and technical activity above a certain threshold in the importing country. A sizeable spread of scientific activity must exist if the selection of new developments is to be relevant to the needs and traditions of the country concerned, and if they are to be fully assimilated. Unless the level of national scientific awareness is sufficient, a country can easily purchase the wrong patents, exploit them ineffectually and innovate its way to obsolescence.

Technological innovation and industry

The technological gap, demystified, proved not to be essentially technological after all. Nor was it a managerial gap, as some had suggested. It reflected instead disparities in the capacity for innovation. The basic process of technological innovation is extremely complex and only began to be understood during the period we are discussing. The existence of a novel idea, derived from research and with the potential for economic exploitation, is the essential starting point for innovation, but is only the beginning of a long process of development. Of many factors involved, a favourable fiscal policy and the availability of risk capital are obviously important; essential, too, is an entrepreneurial environment, management skills, and good relations between management and labour. A market for the new product may exist, or may have to be created so that management skills have to be made available. Behind all this lies the need for propitious general conditions such as an effective education system, while local traditions and psychology have to be taken into account. It is only by understanding these individual variables and their interactions that the innovation process can be orchestrated and mastered.

Within the individual enterprise, the R&D function has developed steadily and, in the great science-based industries such as the chemical, pharmaceutical and electronic sectors, accounts for a high proportion of costs. At the beginning of the period under discussion the relationship between R&D and top management was often unsatisfactory. The research department was too easily isolated from general corporate policy and tended to develop the NIH (not invented here) complex, acting as a barrier to innovations purchased from outside. The firm can expect to develop only a proportion of its products and processes from its own domestic research, yet to rely on purchased technology can be dangerous unless a critical level of scientific awareness exists, to allow the firm to scan the world and select the technology that is best and most relevant to its needs. This awareness is likely to be acquired only through a critical mass of internal research capacity. Articulation of R&D with other elements of company activity is essential. In most cases successful innovation depends essentially on entrepreneurial capacity at the corporate centre; this has to be exercised in close relationship with the skills and imagination of the research department.

Industry is one of the main poles of power in all the developed countries. Industrial firms and especially the transnationals are also major forces in international relations and in the transfer of technology. In the less developed countries, they are needed, welcomed and feared. This is inevitable, since their power and resources are often much greater than those of the countries in which they operate. The large, modern, technology-dominated corporation is undoubtedly extremely powerful, but for this very reason it is in a position to introduce new technologies efficiently to the developed countries. In advanced countries, with important defence commitments, much of the new technology is developed within industry, by contracts with military, space and nuclear agencies of the government. Thus the firms and agencies have a common interest in generating ever more sophisticated equipment. Many fear that the military-industrial complex can evade the control of parliaments and that its enormous power could have sinister consequences. President Eisenhower's valedictory speech warning of the dangers of the military-industrial complex may still be valid today.

Research, development and the arms race

Most of the post-war period has been dominated by the Cold War, which gave rise to an arms race of monstrous dimensions in which military R&D played an essential role. We have argued that technological development arising from discoveries in research laboratories has been one of the main driving forces of economic growth and industrial progress. The importance of such innovation in the military fields has been staggeringly great. It has led to a level of sophistication in military technology said to be several times as great as that in civil development. If there were no military R&D, few new weapons would be invented and improvements in conventional arms would be quite marginal. As it is, research has transformed the nature of war and increased its danger and the extent of potential destruction by several orders of magnitude. The cause of war and the motivations that trigger it appear to be unchanged over the millennia, as apparently does the wisdom that orders our affairs, but the power of the weapons in the hands of aggressors has multiplied a million-fold.

Military research has produced innumerable weapons and systems of high sophistication which impel the arms race and maintain its momentum. The R&D projects of today become the deadly weapons of tomorrow in an escalation of armaments in which technological sophistication is more important than the sheer volume of hardware. Technological advances in the armaments of one of the superpowers were immediately countered by improvements in the strike capacity of the other, in an endless chain of action and reaction. Technological superiority thus became the main preoccupation of the two superpowers in a race of technological complexity and uncertainty which demonstrated the fallacy that military equilibrium could be perceived or maintained.

The finance devoted by governments for military research-and-development was, in its heyday, a large proportion of their total R&D budgets; in the case of the United States and the USSR, more than half. Furthermore, the research input of products is much greater in the civil sectors. The Stockholm International Peace Research Institute (SIPRI) estimated in the 1970s that the average advanced piece of military equipment was about twenty times more research-intensive than any item on the civilian market.

Information about the number of scientists and engineers engaged in military R&D is not known accurately. SIPRI estimated that some 400,000 were so occupied in the 1970s and later estimates raised the number to 500,000. This meant that 45 per cent of the world's research scientists were engaged in military pursuits, or, if only physicists and research engineers were counted, over half. The research cadres thus constitute a formidable component of the military-industrial complex and represent a tremendous diversion of the world's brain-power from constructive to destructive activities.

This huge population depends entirely on defence budgets for support. They are, in the main, outside the general international scientific community and, although many of them must be outstandingly brilliant as the devilish success of their products indicates, they are largely anonymous. Being outside the conventional system of science, which depends for its rewards on the free publication of original work and the judgement of peers, their advancement and recognition within their own circles comes from success in devising more and more sophisticated military devices and assuring their adoption. They are thus a critical group for the continuation of the arms race. Thus scientists join with defence industries and ministries and armament salesmen to form a scientific-bureaucratic-military-industrial complex whose interest is to maintain military budgets, agitating for the development or ever more complex and costly technology for military purposes. The power of this group has appeared so strong as to be virtually irresistible. Has the ending of the Cold War and the substantial disarmament that followed it eroded the power of this complex? We do not yet know. Kenneth Boulding has pointed out one interesting paradox of Cold War behaviour, namely that the defence organizations of the two superpowers, with their driving force of R&D, were actually in de facto co-operation with each other and in competition with their own civilian populations. Thus any increase in the American research budget almost automatically caused an extension of the corresponding Soviet complex and vice versa, while each such extension was an extra burden and danger to the two civilian populations.

Policies for science and technology

The great post-war surge of research funding saw a burgeoning of activity in every sector; in industry, medicine and agriculture. Nearly every department of government discovered a need for research, or at least for scientific advice. Government laboratories proliferated in many countries, including some for new fields such as nuclear power. Also, with a large extension of university education, postgraduate research grew and flourished. The phenomenon of 'big science' appeared; fundamental research which demanded facilities such as high-energy particle accelerators or radio-telescopes which were very costly, well beyond the possibilities of the universities, and for which new methods of funding and organization had to be found. The prewar national structures for R&D were patently inadequate and discussions were taking place everywhere as to what structural changes were necessary, how a comprehensive approach to the total scientific effort of the nation could be elaborated, and how resources should be allocated to support promise and creativity in pursuit of prosperity and the well-being of society. As we have seen, the first national body for science policy was established in London in 1947, the Advisory Council on Scientific Policy (ACSP). Its first annual report outlined coherent institutional proposals for the government's civil science activities that were agreed in principle and largely implemented. In the United States, where the wartime research system had been so successful, several years had to pass before the National Science Foundation came in to being and it was only in the Eisenhower administration that a council for science was set up, chaired by a science adviser to the President, with a broad mandate for the working out of the peacetime strategy. In France, the national Council for Scientific Research (CNRS) was set up and President De Gaulle had his own science adviser. Corresponding arrangements soon appeared in the other European countries and elsewhere.

In the early days, the main preoccupations of the new structures was to devise mechanisms for resource allocation. Initially the scientists were very suspicious of these moves, fearing a bureaucratization of research which would endanger its creativity. Soon, however, professors of high-energy physics realized that it was impossible to expect that each could be given a cyclotron or a synchrotron, and that an intelligent match between resource distribution and research promise could not be made without the advice of the scientists themselves. So the judgement-of-peers principle was accepted. With regard to the financing of 'big science', it was soon found that this was beyond the means of many of the individual countries and would be in any case wasteful. This resulted in the creation of international research facilities, such as CERN in Geneva. Discussion of the many types of international co-operation that have sprung up since the war are beyond the scope of this work, as is an account of the difficulties countries have experienced in creating the ideal structure to include science policy advice, fundamental research, support for agriculture and industry and governments' own research activities. No generalized principles of science structuring and articulation have emerged or are likely to emerge in view of the very wide range of national traditions and political approaches. In many instances the national structures for science have been changed several times in the short history of science policy.

We have been discussing mainly policies for science. On the other side of the coin is science for policy: namely, how a systematic use of the scientific method can help in the construction of policies in other sectors of national activity, foresee some of the consequences of new policies, and avoid counter-intuitive mistakes. Studies of the nature of the innovative process have already proved useful in this sense and there are no social problems that warrant a deeper and more systematic investigation. The application of systems analysis and the devising of computerized dynamic models, which allow different options and scenarios to be explored, hold much promise. With the great increase in the rate of change in so many fields there is a growing recognition of the need for deeper insights into the relations between various sectors, which until now have been considered separately – as, for example, between energy, environment, demographic change and food security. In this new sense, science policy can be regarded as a policy for the mastery of technical and social knowledge and, more widely still, for the generation of new knowledge required by man for the better understanding of himself, his societies and the universe.

New actors began to appear on the political stage with the arising of innumerable non-governmental organizations (NGOs) and the collective views of scientists were often put forward to help in the evolution of official policies, as in the reports by the National Academy in Washington on subjects such as global warming and population. This was also the case internationally where, even during the Cold War, scientists from both sides through the Pugwash Movement did much to inform the political bodies of the true nature of the issues and contributed to the non-proliferation debate. A further considerable impact was also achieved by the report to the Club of Rome, *The Limits to Growth*, which triggered a worldwide debate.

During the period we are discussing, public awareness of science and its applications increased greatly. From seeing science as a mystique to regarding science as the basis of contemporary society was an enormous step. Negative as well as positive consequences of its applications were quickly recognized. Public disquiet was vociferously expressed through the many 'peace movements', advocating nuclear disarmament, protesting against nuclear submarines and opposing the construction of atomic power stations. These movements attracted many scientists. Although at times tinged with fanaticism, they had considerable influence on official policy. We need only cite the referendum that led to the cancellation of nuclear power in Sweden.

A similar situation arose with regard to environmental issues, concern for which has given rise to innumerable study and protest groups. This concern has penetrated the political mainstream in many countries with the appearance of 'green' parties which have had some success in national elections. While single-issue parties are unlikely ever to have a permanent and major role in national politics, the greens have certainly forced the traditional parties, rather belatedly, to cobble together responsible environmental policies. The same is true of industry.

Science, power and the developing countries

It is generally accepted that development in the Third World depends essentially on the creation in each country, or at least groups of countries, of an endogenous capacity for science and technology. This was the theme of the United Nations Conference on Science and Technology for Development (1979), but there has been disappointingly little progress since. This may be due partly to the world economic recession and the diversion of donor interest to the problems of Eastern Europe, but the climate for scientific and technological development is not propitious. While much lip-service is paid to the importance of science, little is done. Science as a career is not fashionable, most of the élites in developing countries being lawyer dominated and inaccessible to scientific advice.

In contrast to this slow diffusion of science and technology into most Third World countries and hence lack of development, the success of the newly industrialized countries of South-East Asia has been a dramatic example of the power of science to stimulate economic development – broadening and deepening of the educational base, creating an infrastructure of R&D and technical skills followed by the purchase of selected foreign patents. The results have been spectacular. This again is an example of how, in the global geography of power, science is playing a fundamental role.

The Industrial Revolution was triggered off by empirical improvements to the steam engine – that is, by invention. The post-industrial, information or service society, as it is variously called, is derived essentially from research discoveries in solid-state physics. Future changes in lifestyle and society will likewise come from discoveries in the laboratory, and will likewise modify the nature of society. Much concern will have to be given to ensure that the new possibilities are exploited in the best interests of society and not merely of individuals.

When it was announced that the OECD intended to convene the first ever international meeting of ministers of science in October 1963, the then Netherlands Minister of Education (later Prime Minister) hurried to Paris to try to persuade the Secretary General to cancel the meeting. 'Science', he argued, 'is an aspect of knowledge, and its policies are an integral part of cultural policy. To discuss them in an economic environment would amount to prostitution.' Today, the pendulum has swung in the opposite direction. Research tends to be mercantilized and projects, even of basic science, judged according to their expected yield as an investment. Much basic research is indeed carried out in university laboratories, but much of it is financed by industry, attempting to push back the frontiers of knowledge at particular places where a deeper understanding is required for a specific application. This is excellent so long as sufficient funds are available for free-choice research, from which the breakthrough knowledge for the future is likely to arise.

At one stage in Canada in the 1960s a change of the science structures was being made and a minister appointed at the centre of government with a broad mandate to co-ordinate scientific activities across the whole spectrum of government interest. In announcing the new appointment a solemn pronouncement was made that 'knowledge is power; science is power'. It soon became clear that the new minister had, in fact little or no power at all: the existing power structure quickly saw to that. Political and administrative power could regard scientific power with contempt.

We repeat what was said at the outset; our world and our societies have been shaped by an enormously successful technology based on the discoveries of science and they are maintained by this. Science can thus be regarded as the power that changed the world for good and for evil. But we can regard science rather as the agent of these changes, in the hands of men and women of power who were able to seize and exploit it. Scientific research is thus the breeder reactor of power, continually rendering more powerful those who already possess power.

The birth of a counter-culture: from Kropotkin to Sakharov

Franco Ferrarotti

A curious paradox must be faced at the very outset: nothing is perhaps more culturally dependent than the various counter-culture movements. In other words, writing and working against the prevailing culture amounts to an essentially cultural enterprise. Culture seems to be inescapable. A supposedly dying, declining or simply inadequate or 'wrong' culture is usually criticized and opposed by a culture keen to take its place. To use the famous formulation of Vilfredo Pareto, one could say that the whole of human history is nothing but the continuing struggle of a 'repressed' élite against a power élite through a variety of means which can be summarized and symbolized, as it were, by two animals: the lion and the fox, that is to say, force and shrewdness.

Taking a bird's-eye view of counter-cultural movements, we can discern three main types. The first is a global approach, on the basis of which counter-culture coincides with a total revolution, involving institutions, governments, power structures and individuals. The second, by contrast, implies an essentially individualistic stand, to be conceived as a self-cultivating attempt in an alien or hostile world, bordering on some sort of self-exile in an environment dominated by an impersonal mechanistic logic and by large-scale industrial machinery. The third seeks a reconciliation of humanistic values and culture with machines and technical progress, especially with electronic devices, to the point of inventing the new, programmatic word 'cyberpunk'.

The first type is most clearly seen in the 200-year-old anarchist movement. Interestingly enough, such movements originate and develop mostly in countries characterized by a backward economy and a large illiterate peasantry. Tsarist Russia and Catholic Mediterranean countries are its hotbed. No matter how strange it might seem, anarchist thinkers, while dreaming of a 'powerless society', are in fact obsessed by power and its prerogatives.

The official founding father of anarchism, Peter Alexeivich Kropotkin, a Russian prince, trained first as scientist (he was a geographer). In 1872, he visited Switzerland and in Geneva joined the Marxist International Workingmen's Association. He found it, needless to say, too conservative. Like many anarchist agitators, his life was marked by exceptional mobility: Russia, then Switzerland, Paris, England, the mountains of Savoy, and finally, like Marx and Engels, peaceful and tolerant London, a safe haven and a permanent residence. Perhaps the gentle atmosphere of London, in a political and social sense, was not without some positive effect on the mental climate of Kropotkin who, in his last years, became more and more interested in 'mutual aid' and in writing up his thinking about agricultural and geographical subjects.

Just as the seekers after a counter-culture are themselves creatures of that culture, so, too, is their relationship to power. The constant struggle against power and its cultural defenders is likely to result finally in a mirror-like counter-power, without ever escaping from its logic. It seems to be an exquisite case of contamination through contact. Fighting power effectively implies the acceptance, at least for practical purposes, of its *modus operandi.* The only alternative to a risky power struggle seems to be renunciation, pure and simple, with its inevitable consequence: the withdrawal from the external world into the inner soul of the individual. The history of anarchism provides some excellent examples of such withdrawal, not to mention the tradition of the ascetic hermits during the tumultuous transition from the last years of the Roman Empire to the early stage of the Middle Ages. The case of St Jerome, who retired from his native Dalmatia to a cave near Bethlehem in order to escape temptations, but also to translate the Bible into Latin, is well known and more than once celebrated by the Italian Renaissance painters. The situation is, however, different when we come to modern counter-cultural movements. Here the withdrawal does not have, as it were, a superior justification. It is not a precondition to opening a special personal avenue for communicating with God; rather, it is a rejection of the external world in order to achieve peace and a clear conscience. Instead of denying the individual in front of Divinity, it is a rejection of the world in order to be able to know and to appreciate the value of the individual *per se*, without the need of transcending him.[1]

During the late 1960s, a powerful and vocal counter-culture developed among young people in the Western world, particularly among university students. The causal or conditioning matrix of the movement

1. In this connection, and for a general survey, see Karl J. Weintraub, *The Value of the Individual*, Chicago.

varied, ranging from the opposition to the Viet Nam war in the United States, to the frustration bred by a lack of political opposition in the Federal Republic of Germany, the anti-paternalistic feeling in De Gaulle's France and the 'baronial', authoritarian atmosphere in the universities and in society at large in Italy. Different as they were in their proximate causes, the various youth and students' movements coincided in so far as certain broad features are concerned.

First, they all represented a radical protest against the 'Establishment', that is to say, against the adult world as organized and ruled by their parents and by those professors in the education system who were seen as father-figures. Second, they rejected the prevailing system of thought and system of production and consumption, which was seen as being soulless, purely mechanistic and impersonal, and exclusively geared to material wealth and greed. Finally, they saw traditional values as being empty and aimed instead at authenticity and spontaneity, believed to be the only acceptable source of moral obligation and capable of returning to man that essential 'plasticity' and 'multivalence' that had supposedly been taken away from him, reducing him to a pathetic 'one-dimensional man'.[2]

Contrary to what was expected by some social analysts, such as Edgar Morin and Alain Touraine,[3] the counter-culture movement of

2. The reference is here made to the famous book by Herbert Marcuse, *The One-dimensional Man*, (London, 1964), a genuine *livre de chevet* for the 1968 generation. Needless to say, the myth of 'total authenticity' had already been fully expounded by Max Stirner, as we have seen, approximately a century earlier.
3. See E. Morin, *La Brèche*, Paris, 1970; A. Touraine, *Ce n'est qu'un début – Mai '68 ou le communisme utopique*, Paris, 1970.

1968 did not amount to a social and political revolution. Although intermittently inspired by the 'Cultural Revolution' of the Chinese Red Guards against their own bureaucratization – the 'revolution within the revolution' was the catchy definition offered by Régis Debray – the 1968 movement could never overcome the fact that it was mostly made up of urban middle-class students who could support themselves on the monthly cheque received from their families. It would be unfair, perhaps, to reduce the whole movement to a purely psychological question of frustration derived from protracted dependency on parental, scholastic or political authority, nor could this be demonstrated empirically. I submit that there were genuine objective problems that went beyond and could not be easily solved in terms of simplistic 'generation gap' theory. It is true, however, that the movement of 1968, for all its vocal propositions, was not able to last or succeed in conducting a cold, realistic analysis of the power structures it intended to attack and eventually demolish. In fact, its final result consisted, paradoxically, in a reinforcement of vested interests and in general of the existing establishment.[4]

From the very beginning of the movement it was easy to foresee that the violent methods that the students resorted to would have had a boomerang effect on general public opinion – in other words, that the effects of the 1968 movement would extend well beyond the academic world. Although the political aim and the acquisition of power in the name of counter-power or no-power were loudly voiced, it was clear at the very

4. In this respect, see F. Ferrarotti, *L'Italia in bilico* (Bari, 1991), passim, but especially the Appendix.

outset that politics would not be the only field of the movement, nor perhaps the most important one. In this respect, the dialectic between young and adults was early caught by an astute observer:

> Many students exhibited their renunciation of the culture of their parents through innovative forms of dress, hairstyles, use of drugs, obscene language, and more widespread willingness to have a variety of sexual experiences. . . . And many adults who could accept revolutionary politics as the idealism of young people who would grow more conservative as they grew older, a frequently made observation about student radicals in many countries, found themselves horrified by the new morality of a renunciatory segment of the young.[5]

It is, however, difficult to deny that the 1968 movement had some links with the individualistic denunciation of machine-based civilization by thinkers such as Thoreau. Nothing and nobody seem further from the aloof anarchist of *Walden* than the obstreperous, rambunctious and at times cheerfully joyous movement of the students marching, for instance, along the Boulevard Saint-Germain in Paris. A common thread, however, becomes visible when they assert that imagination should seize power, or claim to invent a new way to manufacture motor-cars or finally pretend to have a child-care centre on the premises of the Sorbonne. From a safe distance, it is impossible to deny some sort of kinship or at least intermittent congeniality between the movement of 1968 and the antipathy towards machines during the early phases of the Industrial Revolution. It was both a popular revolt

5. See S. M. Lipset, *Rebellion in the University*, p. xvi, London, 1972.

against machinery as a labour-saving device, and an aristocratic pose, such as that adopted by John Ruskin, in *Unto this Last*: that is, a polemical feeling against machinery by people who, because of their privileged station in life, could not possibly have any familiarity with industrial machines.

I have dealt extensively elsewhere with anti-machine feeling as a formal protest.[6] In its crudest form, this protest is traditionally called Luddism. It takes its name from an English worker, Ned Ludd, who, failing to win a dispute with the owners of a shoe factory, vented his anger on the factory's machines. The first group of Luddites made their appearance in Nottingham in 1811, and a deep vein of anti-machine feeling has generally distinguished the early phases of the modern workers' movement. Their machine-breaking fury needs no ideological or doctrinal justifications. It explodes with the natural and violent readiness of legitimate defence. The machine condemns the worker to unemployment and poverty. Furthermore, with mechanization the impersonal and anonymous discipline of 'work rules' takes the place of paternalism, which, albeit authoritarian, is still human. The old-fashioned employer, who chose the best workers from the most lively boys of his parish, is succeeded by the university-educated engineer.

The polemical reaction against science and scientific reasoning is nothing new. When small industries, developed in the old regime as a sort of complementary activity to large-scale landowning, give way to the first accumulation of capital, which makes large-scale mechanization

6. See F. Ferrarotti, *The Myth of Inevitable Progress*, pp. 26–7, New York, 1985.

technically possible and economically convenient, the consequences for the whole way of life are immediate and profound. Here a more pervasive and anti-machine feeling arises and develops into two different directions, which could perhaps be called the 'mystique of craftsmanship' and the rejection of the modern scientific world. The 'mystique of craftsmanship' is not anti-technology in principle; it accepts technical progress with reservations. It recognizes the advantage of technical progress but fears that the price is too high. It proposes decentralization; it preaches against the mania for the colossal and size in all its aspects. Its most coherent historical representative is the French libertarian socialist or, perhaps more precisely, anarchist, Pierre-Joseph Proudhon.

Other groups are more extreme and collide head-on with the modern world as the world in which, to use the famous title of Siegfried Gideon, 'mechanization takes command'. Despite the intellectual refinement of these groups, we can easily detect in them overtones of medieval and decadent nostalgia. The romantic German poet, Novalis, had already noticed the damage that, in his judgement, science was doing to the 'sense of the invisible'. But even sophisticated modern writers such as Dwight MacDonald seem to share the romantic tendency to downgrade the scientific mind:

> We Americans are hunters rather than artists, men of action rather than of thought or feeling. Our chief contribution to philosophy is pragmatism . . .; technique rather than theory distinguishes our science; our homes, our cities, our landscapes are designed for profit or practicality but not generally for beauty; we think it odd that a man should devote his life to inducing children to breakfast on *Crunchies* instead of *Krispies*; our

> scholars are strong on research, weak on interpreting the masses of data they collect.[7]

This is, according to MacDonald, the 'triumph of the fact' that is brought about by a dominant scientific attitude.

It seems clear that we have here an emblematic example of the confusion between science and its popular caricature – 'scientism'. This pathetic error, though widespread, regards science as a kind of mysterious and all-powerful magic art, and is due, at least in part, to scientific propagandists who tend to see in science and in scientific research the cure for all the evils of the world. From such commonplaces the scientist emerges as a new saint, the saint of the industrial age of mass technology. Science itself is conceived and presented as a new religion. It is seriously stated that science, given its head, is not just cold mechanical efficiency; its attitude is tolerant, friendly and human. It has already become – it is maintained – the dominant inspiration of human culture, so that modern poetry, painting and architecture derive their most constructive ideas from scientific thought. This mood goes as far as ignoring simple facts that are connected with scientists' responsibility. Twenty years before C. P. Snow, the celebrated author of *The Two Cultures and the Scientific Revolution* (London, 1962), C. H. Waddington took the love for peace on the part of the scientists for granted:

> Far from considering war as the highest activity of man . . . scientists even more than most people in this country [the United Kingdom], find it a damnable, if unavoidable, interruption of their serious business. For science is concerned

7. See D. MacDonald, *Against the American Grain*, New York, 1952.

> with man's real evolutionary advance, with giving him the mastery over things which will enable him to translate his wishes and aspirations into fact.[8]

Too bad, one may add, that scientists have also provided politicians with the most powerful and deadly means of mass destruction.

Quite unexpectedly, it was among the most advanced scientists, particularly in nuclear physics, that the reaction against the practice of scientific research as a neutral, purely instrumental activity – a kind of perfection without purpose – was formed and then, little by little, developed. The most eloquent and telling example of a positive counter-culture movement is not, therefore, to be looked for among vociferous 'revolutionary' students, but among dedicated researchers and top-level scientists, Robert Oppenheimer and Andrei Sakharov. These scientists and their followers have created through the years since the end of the Second World War an 'invisible college'. The function of this college has been to work in the scientific laboratories as well as with public opinion at large in order to turn the resources of nuclear energy away from war to peaceful activities and practical application in various fields, from medicine to electricity production. Even in the earliest days of the

8. See C. H. Waddington, *The Scientific Attitude*, p. 160, London, 1941. C. P. Snow was also preceded by scholars such as James Harvey Robinson (*The Humanizing of Knowledge*, p. 81, New York, 1923): 'Those who devote their professional lives to literature can probably look back to the vain efforts of their awkward teachers of science to adorn them with a dab or two from Science's iris-hued brush and their lasting resentment at the attempt to discolor their poetic or romantic souls. In later life they compensate themselves for their gross ignorance of natural processes by declaring, as Mr. Chesterton does, that science is irrelevant to our deeper lives; or as Brunetière did, that science is bankrupt.'

Bulletin of Atomic Scientists, the scope of such initiatives was extraordinarily broad. To the groups of dissenting scientists, centring around the *Bulletin* not only physicists but also social analysts such as Edward A. Shils and David Riesman were drawn.

Nuclear physicists have undoubtedly influenced public opinion on at least the technical aspects of the peaceful uses of the atom. But in some cases this contribution has gone further, turning into a political struggle in the proper sense. While Leonid Brezhnev was in power in the former Soviet Union, Andrei Sakharov and his wife Helena Bonner were forced to leave Moscow for exile in Gorki, where their visibility, especially in the eyes of foreign public opinion through newspapers and television networks, was dramatically diminished. In other cases, opposition to the use of nuclear energy for military purposes was less conspicuous but nonetheless constant and also courageous. Robert J. Oppenheimer, leading physicist of the Manhattan Project during the war, refused to support the construction of the hydrogen bomb, citing technical difficulties. In doing so, he was in fact hiding his personal moral scruples behind certain technicalities. He went to the point of resigning his important government post. A man less disturbed by moral issues, Dr Edward Teller, would eventually build the hydrogen bomb.

Given the crucial importance of their work, it is no wonder that the political loyalty of the whole category of nuclear scientists has often been questioned. The situation was made worse by inevitable internal dissension among scientists. Only the naïve would in fact imagine that scientific progress unfolds according to a smooth, harmonious and self-correcting logic. The autobiography of Emilio Segré, an important collaborator of Enrico Fermi since the earlier days in Rome, proves once more that cold-blooded and self-controlled scientists are also passionate human beings and can have sentiments of warm sympathy as well as of profound dislike for others.[9]

Segré flatly states in his book that the relationships between scientists, especially physicists, are usually 'veiled by a reciprocal doubt or suspicion'. Segré is particularly bitter and, it seems to me, unfair towards Robert J. Oppenheimer, whom he found not only unsympathetic but also 'slightly ridiculous'. Naturally, his basis of comparison was the towering figure of Enrico Fermi, 'a man of great solidity, whose simplicity was in stark contrast with the elaborate and complicated erudition of Oppenheimer'. Segré has also something to say about the Oppenheimer group, whose average cultural level, in his opinion, 'was somewhat below the level of European high-school students'. It would be wrong, however, to reduce the internal frictions among these scientists to purely personal idiosyncrasies or psychological feelings. The very fact that nuclear physicists were dealing with the construction of secret, and powerful, weapons would expose them to the traps and temptations of the spy rings from both East and West. The recent book by Pavel Sudoplatov and Anatoli Sudoplatov[10] is, in this respect, an important, though controversial, contribution. Its main thesis is that scientists such as Oppenheimer, Fermi, Szilard and Bohr were manipulated and used by Soviet intelligence. While this may be a gross exaggeration, it is well known, on the other hand, that at least some prominent scientists in the West (Fuchs, Pontecorvo) believed that in passing atomic secrets to the

9. See E. Segré, *A Mind Always in Motion*, Berkeley, Calif., University of California Press, 1994.
10. See P. Sudoplatov and A. Sudoplatov, *Special Tasks*, New York, 1994.

Soviet Union, they would help to keep peace in the world through a nuclear balance between East and West. Indeed, in the years immediately before the Cold War when the Western Allies still thought of the Soviet Union as a reliable partner in the struggle against Nazism and fascism, disclosing classified information could, I submit, be regarded as much as a counter-culture move, more or less ideologically inspired, than a spy initiative.[11]

In this connection one cannot help admiring the sober attitude of Andrei Sakharov. Confronted with immediate danger of persecution by the Soviet authorities, Sakharov was nevertheless adamant in defending the cause of peace and international understanding. What seems to me, at this point, most admirable is the sense of measure and balanced judgement he showed time and again. Devoted as he was to the cause of disarmament – even of unilateral disarmament if necessary – he never gave up the idea of nuclear testing as the only means of assessing scientifically the impact of the new weapon technology by either side. At the Forum for a Nuclear Free World, held in Moscow in February 1987, Sakharov is reported to have said:

> The question of nuclear testing is not critical for the restraint of the arms race. The issue of nuclear testing, in my opinion, is of minor, secondary importance in comparison with the other military, technical, political and diplomatic problems involved in preventing thermonuclear calamity. . . . I believe that eliminating the issue of a comprehensive nuclear test ban will facilitate negotiations on more urgent problems of disarmament.'[12]

11. See the interesting polemical exchange, 'Were the Atom Scientists Spies?', *New York Review of Books*, 22 September 1994, pp. 72–4.
12. See A. Sakharov, 'Of Arms and Reforms', *Time Magazine*, 16 March 1987.

Thus, we have come a long way. In approximately two centuries, from the anarchist Kropotkin to the scientist Sakharov, the counter-culture movement has come of age. It is no longer to be confused with the grandiose, but highly unrealistic, dream of a world liberated through a cathartic revolution, but it rather emerges as a piecemeal transformation which is constantly under the control of the community, in order to strike the best connection between what is desirable and what is possible.

Genetics, ethics and power

Ideological deviation and ethical protection

Gérard Huber

Nowadays there is no argument about the immense benefits that mankind derives from biology, psychology and medicine. In medicine, the revolution that began with the sulphonamides in 1937 gave man, in the words of J. Bernard, the power after thousands of years of helplessness to conquer what had long been fatal diseases responsible for large-scale mortality.[1] Today gene therapy is opening the way to techniques for treating disorders controllable by genetic means and until now beyond the reach of scientific know-how. Lastly, there are the achievements of neurobiology and psychoanalysis and the light they throw on neuro-synaptic mechanisms and on affects, impulses and representations, so that the image of man is now very different from what it was little more than a century ago. Understanding better in order to live better could aptly summarize the basic contribution that science has made to all our lives. Modern societies now recognize that science has its rights. Unfortunately science can also be used by some people to justify order

1. J. Bernard, *De la biologie à l'éthique*, p. 22, Paris, Buchet/Chastel, 1990.

and hierarchies, as if their own political rules were not sufficient. Particularly apparent in authoritarian and totalitarian states, this may also arise in democratic societies.

For many centuries it was the physical understanding of the world – and even the universe – that was used to justify the proclamations of official truths concerning the social hierarchy and its translation in terms of power. To go back to the works of Antiquity, there is Hesiod's myth of the metallic races and the adaptation that Plato made of it and set out in one of the dialogues between Socrates and Glaucus. Then there are the analogies drawn between the hierarchies of temporal power and those of the celestial world, of which in Ptolemy's model the earth formed the centre with an ordered arrangement of celestial bodies gravitating around it.[2] Closer to our time there is the Catholic Church's obscurantist theocracy, its justification of the social order in terms of the Creation and its refusal to accept a universe built according to the physical laws discovered by Galileo. It is, incidentally, only very recently that the Church has officially acknowledged its error.

At the same time, however, official conceptions of the world and of man were also borrowed from a vague biology only perceivable by the senses. It was not, however, until the dawn of the seventeenth century that speculation stemming from the very heart of biology made its first contribution to the ideological systems of political authorities and social reformers anxious to justify racial and social hierarchies. Thus it was that Buffon's ideas about naturalism had such a decisive influence on European Latin or Catholic thinking about social conventions,

2. S. J. Gould, *La Mal-mesure de l'homme*, p. 26, Paris, Le Livre de Poche, 1983.

while Linnaeus had a comparable effect on Protestant thinking.[3] Gradually a 'theory of degenerations' began to emerge. Buffon started from the 'geobiological' idea that a human nature existed which steadily degenerated as man spread over the earth and came closer to the pole or the equator. Linnaeus put together a binary classification to explain a hierarchy reflecting the various modes of government. The classifications of the eighteenth century identified four or five human races defined by colour of skin and pseudo-moral or -mental qualities. David Hume, philosopher and Under-Secretary of State in the British colonial office in 1766, believed in the separate creation theory and the inborn inferiority of non-Whites. In the United States it was taken as self-evident that the Blacks were a race and that this race was inferior to that of the Whites. Disagreement between group doctrines was mainly a matter of one side (the doves) accepting and the other (the hawks) refusing to accept that libertarian rights applied to Blacks as well as to Whites.

In the nineteenth century the next stage was what is called 'social Darwinism' (no doubt wrongly, because it is doubtful whether Darwin would have gone along with the tenets of this ideology), which is simply the use of the theories of natural selection and their projection into a scheme of social selection and hierarchization. Thereafter, it was on biology that the powers of the twentieth century – particularly the dictatorships but not they alone – tried openly to found their ideology, whether racial or social or both.

3. L. Poliakov, *La science face au racisme*, pp. 70 et seq., Paris, Éditions Complexe, 1986 ('Le genre humain' series).

In the first part of this chapter we try to show, taking as illustrations the use of mass psychometry in the United States, the misuse of genetics by the Nazi authorities of the Third Reich and Lysenkoism in the Soviet Union, how discriminatory ideology in the recent past has sought to base itself on biology. In the second part we look at new scientific advances to which new discriminatory ideologies might be tempted to turn and the protection against them afforded by the democracies.

American democracy and the use of mass psychometry

Psychometry is an attempt to measure the development of intelligence. It was conceived when Binet (1857–1911) appropriated the methods of Broca's 'craniometry' and Lombroso's anatomical stigmatization[4] and used them to fashion a psychological tool in the form of mental tests. Developed in France to identify the main obstacles to learning in low achievers at school, the method took the form of a scale of tests allocated in accordance with the tasks that each age group was capable of. Each test was scored, enabling a general assessment to be made of each child's ability to answer. A 'mental age' was assigned to each test. The general level was assessed after working out the difference between the 'mental age' and the chronological age. In 1912 the psychologist Stern proposed, instead, that the 'mental age' be divided by the chronological age and this was the origin of the Intelligence Quotient, or IQ.

Gradually a trend emerged, particularly in the United States, to see IQ – though it depended on psychological methods – in terms of biological

4. Gould, op. cit., p. 180.

heredity. The theories of H. H. Goddard, L. R. Terman and R. M. Yerkes were applied in the armed forces, in business and in schools, and a whole swarm of discriminatory arguments developed which were based on the assumption of a 'heredity of intelligence', an ideology that influenced United States Government policy so much that resolutions passed after debate in Congress in 1924 were founded on supposedly scientific racism. The same ideology was partly responsible for the passing of the well-known Immigration Restriction Act, which laid down intake quotas based on the results that representatives of certain countries obtained in tests distributed in the army and elsewhere.

The ideological use of psychometric tests led to unethical behaviour. In the State of Virginia a law only repealed in 1972 legalized the sterilization of adults found to be mentally defective, that is, scoring a mental age of 7, 8 or 9 on the Stanford-Binet intelligence scale. The same applied in Indiana and some thirty other states. In 1927 a Supreme Court ruling declared that it would be better for all if society were able to prevent the patently incapable from reproducing rather than have to execute degenerate offspring for murder or to see them die of hunger because they were idiots.

National Socialism and the abuse of genetics

In *Mein Kampf*, Hitler declared that without a biological basis and a biological purpose policy was totally blind. Nowadays we are constantly referring to the concept of diversity, but we should not forget that this same principle was also upheld in Nazi ideology, clearly summarized in the words 'the diversity of human beings has a biological basis'.[5] The

5. Benno-Muller-Hill, *Science nazie, science de mort*, p. 15, Paris, Éditions Odile Jacob, 1989.

vital thing, however, is not so much to recognize diversity as to accept both diversity and equality. But the concept of equality was contested by fascist ideology, founded as it was on the principle of fundamental inequality according to which some deserved to live and others did not. Indeed, the purpose of the recruitment of eugenists, psychiatrists, anthropologists and physicians by the state in Germany was to draw up a 'biological/hereditary document'[6] for each citizen (i.e. those considered deserving to live, in other words those Germans who were neither Jews nor gypsies).

The principle of a life that did not deserve to be lived applied across the board to all who were 'different' and operated at five levels: the compulsory sterilization of the 'mentally ill'; the killing of 'insane' children; the killing of insane adults; the holding of the 'insane' in concentration and extermination camps; and the mass murder of mainly Jews and gypsies in these same extermination camps.[7]

The Nazis' basis for justifying the principle that some did not deserve to live was genetics. Their basic idea was that since geneticists, psychiatrists and ethologists understood human genes, the structure of the brain and the behaviour of animals, they could also explain why society was divided into what they considered to be two broad categories of human being – the Aryans for which their dream was the 'new man' and the others, the 'sub-humans'. It was completely wrong. The hereditary and biological nature of psychiatric disorders had never been proved; the very idea of biological transmission of the characteristics of what they

6. Ibid., p. 25.
7. R. J. Lifton *The Nazi Doctors: Medical Killing and the Psychology of Genocide*, New York, Basic Books, 1986.

called a 'race' had no scientific foundation whatsoever. Yet they managed to win the support of German scientists and doctors by promising them a new role in society, that of 'doctor-priest'. This crime against humanity, the mass murder of millions of people purely because of the fact of their birth was therefore aided by the scientists of one of the most cultured nations in the world.[8]

Communism and Lysenko's 'dialectic materialism of nature'

There is more than one way in which the example of Lysenkoism and its relations with Stalinism is significant. It clearly shows than in its nineteenth-century origins, communism wished to be based on a social and historical science claiming to offer politicians a model of 'the good society' and 'the new man', and needed the backing of natural science to support its claims.

In this way Stalinism justified the cruelty of its authority as a necessary transition to the new man, arguing that the dictatorship of the proletariat had a scientific basis, 'proletarian science' conqueror of 'bourgeois science', representative of capitalism. The instrument of that victory was dialectic and historical materialism, a theory of nature and society formulated by Marx and Engels. The October Revolution had proved the historical truth of the theory and it was now left to biology to prove its natural truth. It was for this purpose that Stalin had begun to use Pavlov's conditioned reflexes and had then appointed T. D. Lysenko, a biologist who understood this ideological task as something he could do.

8. A. Frossard, *Le crime contre l'humanité*, Paris, Le Livre de Poche, 1987.

So Lysenko made it his business to make a clean sweep of classical genetics, denying genes and chromosomes any part in hereditary transmission and declaring – experimental genetics notwithstanding – that there was no specific material basis for heredity. A champion of Marxist-Leninist scientistic determinism, Lysenko virulently attacked the chance element in Mendel's laws and mutations and clung to the outdated dogma of inheritance of acquired characteristics.[9] Knowing, too, that the number-one problem in the Soviet Union was food, he moved into agriculture claiming, among other things, that he would quickly bring out new varieties of winter wheat capable of withstanding the cold of Siberia.[10]

The domination of this pseudo-scientific ideology resulted not only in the demolition of classical genetics, whose most representative biologist was Nicolai Vavilov, and the further impoverishment of Soviet agriculture, but also the branding and ruin of all who thought differently. Vavilov, for example, was tried and condemned to death. In the end he was sent to prison instead where he ended his days in awful conditions.

Today we are in danger of embracing a new conception of the 'life that does not deserve to live' principle. Can the democracies protect human dignity and the humanity of every human being against discrimination based on the arguments that ideologists hope to find in biology or medicine?

9. D. Buican, *Lyssenko et le lyssenkisme*, p. 6, Paris, Presses Universitaires de France, 1988, ('Que sais-je' series).
10. Ibid., p. 69.

First it has to be noted that there still is an orthodoxy, a body of ideas, that right-thinking people are assumed to accept without question.[11] At the heart of this body of ideas is the notion of an 'élite', better qualified than the mass of humanity. Some authors would like to reinstate the idea of the heredity of intelligence. Examples are sociologist Charles Murray and the psychologist Richard Herrnstein who attempted to explain the so-called inferiority of Blacks compared with Whites by behavioural genetics and the misuse of IQ testing.[12] Other writers have sought to give substance to the 'élite' myth by drawing on biology and biotechnology. Eugenics, for example, whose history is as old as that of mankind, is the subject of fresh scientific development in the form of a discriminatory ideology of reproduction.

Traditionally two forms of eugenics are recognized, negative and positive. The aim of the former is to limit or even prevent the reproduction of what are considered to be undesirable hereditary traits, while that of the latter is to encourage the reproduction of what are deemed to be 'superior' hereditary traits.[13] Negative eugenics in the past took various forms, ranging from the killing of children judged abnormal to the killing or sterilization of the parents. But it has not vanished from the world. Recently it was revealed in the press that a Swedish Act gave doctors in that country the right to sterilize on the grounds of 'social' or

11. Orwell, quoted by E. Mayr, in *Histoire de la Biologie*, p. 87, Paris, Fayard, 1989.
12. In *The Bell Curve: Intelligence and Class Structure in American Life,* New York, The Free Press, 1994.
13. A. Begin, 'Du choix des procréateurs aux choix des gamètes, sur quelques avatars de l'eugénisme', in G. Huber and P. Jalbert (eds.), *L'Heure du doute,* Paris, John Libbey Eurotext, 1994 ('Ethique et sciences' series).

'racial' health. In the United Kingdom, too, the courts not long ago ordered that a 17-year-old girl be sterilized on the grounds that, for what were described as 'mental reasons', she was incapable of distinguishing between sexuality and reproduction. Some other democratic countries have taken similar measures. A little while ago in Germany, for example, draft legislation made provision for the sterilization of the 'mentally handicapped'.

Positive eugenics, meanwhile, is going through a revolution. For a long time it was a matter of parent selection. In *The Republic* Plato dreamt of totalitarian collective eugenics. Nearer our own time, Tomaso Campanella imagined 'the city of the sun'. These utopian constructions had no need to draw on technology or science. But today advances in biology are resurrecting them and raising difficult questions for policy-makers and social reformers. Contemporary positive eugenics can be defined as the qualitative control of human reproduction. It relates less to parents than to gametes. The fact is that biology can now predict in advance the risks of certain genetic diseases in certain people. But quantitative and qualitative control would merge once societies or governments decided to reduce the number of children born with clinically defined weaknesses that were not amenable to effective treatment.[14]

The illusion here is that the individual can be defined in terms of his genes and other biological characteristics.[15] The belief is that it is

14. MacFarlane Burnet, Nobel Prize-winner 1960, quoted by A. Begin in ibid., p. 156.
15. See G. Huber, *L'Illusion métabiologique*, Paris Presses Universitaires de France, 1994.

possible to find the formula of humanity while it is of unquestionable value to identify sequences of genes for therapeutic purposes, the idea of mapping the genome of all human beings and using this to identify the source of all human characteristics is an obvious example of this belief. The dream of mapping all the neurons, synapses and brains of all human beings is another.

The illusion does not stop at biology or neurobiology. It has crossed over into psychoanalysis, with S. Ferenczi's claim to have laid the foundation of a reconstitution of all the phylo- and ontogenetic stages of the human psyche and, lastly and most recently, into sociobiology where the object is to explain the development of social concepts in terms, first, of growth in the age structures of animal populations and, second, of their genetic construction.[16]

This metabiological illusion is reinforced by the fact that human science has revealed its inapplicability to political thought and its supreme hope, that of defining 'authentic man' or 'the good society', has proved vain. Beyond the indisputable breakthroughs in biology, neurobiology, neuroscience, psychology and psychoanalysis, the metabiological illusion consists in treating the human species as though there are simple scientific explanations for the entire complex range of human emotion and experience. When a scientist fails to object to the (ideological) use of projections of his own scientific hypotheses to explain human and social experience, he runs the risk of implying that he believes in 'best of all worlds' theories.

16. C. Frankel, 'Les enjeux de la sociobiologie', *La science face aux racismes*, Paris, Éd. Complexe, 1986. ('Le genre humain' series).

The protection offered by democracy

The history of societies and that of biological, psychological and medical science today face the challenge of using man's increasing knowledge of man in the interests of the development of each individual. To achieve this, the contemporary conscience has to realize that the primacy of the person and human dignity comes before any other political, economic or social consideration.

Though hierarchical, social and racial categorization were strongly opposed by the humanistic ethics of the Universal Declaration of Human Rights, which relates to the rights of the person or the rights of the human being, the Declaration contains no specific provisions on scientific, biological or medical experimentation. Reference to these had to wait until 1966, with the United Nations International Covenant on Civil and Political Rights.[17] Meanwhile, important recommendations about crimes against humanity were passed by the Nuremberg tribunal in 1947. Those concerning the use of biomedicine were collected together under the heading of the Nuremberg Code, the merit of which is that it condemns any use of biomedical experimentation for purposes of oppression. It also makes general the indispensable need for the 'voluntary consent' of any person agreeing to co-operate in experimentation. To guarantee its democratic application the major international agreements (Helsinki 1964, Tokyo 1975 and Manila 1980) amend the term to 'free and informed consent'. While this guaranteed its application, the

17. C. Byk, *Les progrès de la médecine et de la biologie au regard de la Convention Européenne des Droits de l'Homme,* p. 9, Strasbourg, Editions du Conseil de l'Europe, 1994.

dropping of the word 'voluntary' could be a mistake. The qualification should in my view be widened to 'voluntary, free and informed' consent.

Later, committees on the subject of ethics began to be set up in an increasing number of hospitals or health organizations. In 1983 a new model was established in France, when President François Mitterrand signed a decree setting up a Comité Consultatif National d'Éthique. Since then, in the European Union alone, many national ethics committees have been formed, as in Belgium, Denmark, Italy, Luxembourg, Portugal, etc. In addition, there is the Council of Europe's Steering Committee on Bioethics and the Group of Biotechnology Ethics Counsellors and other special groups created by the Commission of the European Communities. At the international level, UNESCO has recently added an International Bioethics Committee to its structures.

Given the scale of the issues, all relating to applications of biomedical technology, political authorities have had no alternative but to draw up legislation making the legal security of individual persons the first priority. The point is that, unlike dictatorships, democracies have the advantage that there is no revealed dogma or official truth, whether in science or in thought. There are relative truths but even these are highly dependent on societies that are increasingly multicultural because of world travel and long-distance video and audio communications media.

The Council of Europe, for example, has recently produced a draft European Bioethics Convention which will make a useful addition to the European Convention for the Protection of Human Rights and Fundamental Freedoms. And in the six years between 1988 and 1994, France has passed four laws regarding the protection of persons lending themselves to biomedical research, respect for the human body (the body and no longer simply the person now figures in civil legislation), the donation and use of parts and products of the human body, medical aid to procreation, prenatal diagnosis and finally the processing of personal data for purposes of health research.

This cursory review of the parallel development of a 'human rights' tradition and an 'ethics' tradition shows that democracies stand the best chance of finding truly human answers to the difficult questions raised by the contradictions between values and interests, provided they see them in the correct relationship to one another. To give one example, this is the kind of task that the International Institute for Human Rights Studies has taken on by agreeing to become a consultative body on ethics for UNESCO. The task is all the more urgent in that the deregulation of market mechanisms and the widening of the financial, economic and social gap between developed and developing countries may open the door, if we are not careful, to a new form of slavery – the trade in human organs. Another danger lies in the unbridled use of genetic testing and diagnosis (from the stage of the gametes and the ovum up to that of the adult), which would bring into being a new eugenist dialectic of master and slave, open to totally arbitrary application.

Unhappily, while democracies may recognize these dangers and be eager to avoid these new forms of slavery, it depends on a logic that may indeed take ethical, cultural, lay and religious values into account, but does so on the same terms as economic interests and social regulation.

But ethics is not the same kind of criterion as the others. It is as important to human existence as physical and moral security. Ethics will not disappear, even if political authority declares itself to be ethical; it is a vital informal counterweight. To have its effect, this counterbalancing power needs to be permanently engaged on the task of dismantling the myths (perfect child, ideal child, new man, superman, society of excellence, and so on) which idealize man or society and create the impression that the formula for the best of all worlds is near at hand. But this dismantling must not be simply an acceptance of the forces of despair. On the contrary, ethics should be a personal effort which creates hope, and enables one to bring one's acts – public and private – into harmony with one's values at each moment of one's existence, in a research laboratory, a health centre, in places of education, etc. In the future, one of the ways in which ethics and democracy will increasingly assert themselves will be the deepening of humanity's personal and self-narrative dimension in each one of us.

Use and misuse of genetics

Raymond Daudel

For billions of years the dominant factor in the evolution of living species was their adaptation to the environment. Now the knowledge human beings have accumulated has enabled them to reverse the process, to adapt the environment to their needs and, consequently, to influence the evolution of other species. The growth of genetics is opening the door to a third possibility: it enables human beings to produce new species by means of genetic engineering. *Homo sapiens* can even transform himself and become more than ever the 'shepherd of his own destiny'.

As Yves Christen has put it:

> Fearful of entering the age of superhumanity, we are again face to face with ourselves. Will we have the courage to cross the threshold? If so, will we shape our own future? If we do not, the future will condemn us. It is an old problem: should we create standards or submit to them? Feverishly vacillating between anguish and hope, we are like children who wonder whether they dare venture into a magic grotto. It is a terrible, superb moment. Will we have the courage?[1]

1. Yves Christen, *L'Homme bioculturel: de la molécule à la civilisation,* Paris, Éditions du Rocher, 1986.

We know that our cells contain DNA, the molecules that constitute our genome. This contains the information largely responsible for an important part of our physical characteristics and behaviour, so much so that it has been said that a human being's future is already largely present in the egg cell that will give him or her life, that the egg already possesses a memory of the future. A central question immediately arises: How much in human beings is innate and how much acquired? This question is not a banal one, for all the use and misuse that the ideological and political power has made of it.

Much research has been done in an attempt to answer this question. We will mention only two examples, one concerning animals and the other human beings. Take a squirrel and feed it regularly from birth with soft food in a cage devoid of litter, where it can neither see another squirrel hiding food nor try digging for it, or experience hunger. Such a squirrel cannot learn that in times of famine it could take advantage of food it had previously hidden. If, when it is adult, it is given nuts it begins by eating them. When it has satisfied its hunger, it does not lose interest in the nuts but carries one in its mouth to a corner of the cage, begins to scratch, puts the nut on the ground, pushes it with its muzzle and makes movements as if to cover and press it down with its forepaws, though it has not dug a hole. A complete pattern of the squirrel's behaviour is therefore programmed from birth. This is innate behaviour.

As identical twins possess the same genome, the findings of studies of their behaviour have made a significant contribution to study of this subject.[2] Twin sisters, Gladys and Helen, were separated at the age of

2. See, for example, A. Montagu, *L'Hérédité*, Paris, Marabout Université, 1967.

18 months. They did not see each other again until they were 28 years old. Helen was adopted successively by two families. The second adoptive mother gave her a good education. She obtained a higher-education degree, married a cabinet-maker when she was 20 and became a teacher. Gladys was not so lucky: when she was in her third year at school, her adoptive father fell ill. The little girl's education came to an end and she looked after the house. She married a mechanic when she was 21.

The women were physically similar: Helen was 2 cm taller than Gladys and weighed 700 g more. Their hair and eyes were similar. Each had six fillings in her molars and premolars. The difference in their education, that is, in what they had acquired, had above all affected their behaviour. Helen was self-assured, gentle and graceful. Gladys was awkward, ill-at-ease and slow, and lacked grace and charm.

It cannot be deduced from this, however, that our genome does not influence our behaviour. Experiments on rats clearly show the contrary.[3] Researchers have measured an intelligence coefficient in rats by evaluating the speed with which they learn their way in a maze. The males and females that have shown the greatest ability to learn are then separated from the slow ones. Fast rats are mated with each other, as are slow rats. After seven generations, it is clear that two very different sorts of rats have been created: those that after a very short time make no further errors, and far less intelligent ones which learn to avoid error very slowly. There seems to be no doubt, therefore, of the existence of genomes favouring this sort of performance.

3. See, for example, F. Gros, *Les secrets du Gène*, Paris, Odile Jacob, 1986.

The dialectic between nature and nurture certainly begins very early in the life of living beings. Close examination of the eyes of twin daphnids has shown that there are differences in the organization of nerve cells in genetically identical (known as isogenic) individuals. In particular, it has been shown that isogenic daphnids possess in their single eyes exactly the same number of neurons (nerve cells) but that the dendrites of equivalent neurons differ from one daphnid to another. It is these dendrites that establish the contact (synapses) between neurons thus forming the neuronal network of the nervous system.

It is therefore clear that even the neuronal networks of twins' brains are different from one another. A remarkable feature of the development of the human brain is that it continues long after birth. Its weight increases fivefold between birth and adulthood. The great majority of the synapses in the cerebral cortex form after birth and make possible the gradual 'impregnation' of the cerebral tissue by the physical and social environment. The development of the nervous system is therefore accompanied by a growing difference between genetically identical individuals. This difference is beyond the scope of simple genetic determinism.[4]

The important findings obtained by D. Magnusson should also be mentioned.[5] He followed a cohort of 538 Swedes from the ages of 13 to 26 and found that generally it was those who at 13 were both the more aggressive and the more unruly who committed crimes between the ages of 18 and 26. He also showed that these future criminals excreted half

4. J. P. Changeux, *L'homme neuronal*, Paris, Fayard, 1983.
5. R. Daudel and N. Lemaire D'Agaggio (eds.), *Life Sciences and Society*, Elsevier, 1986.

the amount of adrenaline in their urine at the age of 13 than the other children.

This brings us very close to the concerns of sociobiology, which seeks to study the biological bases of social behaviour,[6] and sociobiologists do believe that genes play an essential role in determining such behaviour, some claiming that living organisms are merely the means used by genes to multiply. Like any new science, sociobiology seems to promise a great deal but provokes great controversy as its theories need to be perfected by long experimentation in accordance with accepted usual scientific methods.

6. On this subject, see Y. Christen, *L'heure de la sociobiologie*, Paris, Albin Michel, 1979.

Science at the service of power: the responsibility of scientists

Science and power today and tomorrow

Federico Mayor

C'est parce que les sciences et les techniques sont censées mettre, indifféremment à notre portée des avenirs possibles . . ., qu'il faut que la société humaine décide, au nom de ses valeurs, ce que sera son avenir[1] (Ilya Prigogine and Isabelle Stengers, in *Entre le temps et l'éternité*).

We live in an age where science daily demonstrates its power. The knowledge it has given us has conquered distance, abolished want, reduced disease, and opened doors to understanding the mysteries of nature. As an instrument for discovering 'the truth' – what 'lies behind' – science is without equal. Yet we should be wary of celebrating its triumphs too uncritically, for they have been unevenly distributed. In the midst of plenty, there is want; diseases long understood by doctors still carry off the poor and underprivileged; and in its tidy explanations of natural

1. It is precisely because science and technology are supposed to put within our grasp possible futures that are all equally valid . . . that human society must decide, on the basis of its values, what its future will be.

phenomena, science leaves many people unmoved or even alienated. What science tells us may be true, but it is not the only truth that matters.

The successes of science have made relations between the scientific community and the state far more complex than they once were. No modern society is imaginable without the support of science and technology, and the relationship cuts both ways. Modern science has become too large and too costly to survive for long without subventions from the state. Both sides need each other, but the relationship between science and power is an uneasy one. Scientists want society to support but not to govern science, yet increasingly the public demand is for some form of social control, never more clearly seen than in the ethical debates over the human genome and the manipulation of embryos. Governments, for their part, would like to plan research, to direct it purposively into areas of inquiry that they believe may produce an economic return, an effort that many scientists consider misguided. In the words of Michael Polanyi: 'The pursuit of science can be organized in no other manner than by granting complete independence to all mature scientists . . . the function of public authorities is not to plan research, but only to provide opportunities for its pursuit.'

Science at the service of power

Historians will, I believe, identify the Second World War as a turning point in the relations between science and government. The American president of the time recruited, for example, the first science adviser to the White House. This kind of assignment was something quite new to those who had hitherto earned their living mainly by research or the teaching of science. As a consequence, science helped defeat the enemies of freedom and democracy – enemies who had initiated hostilities in a greater state of military preparedness, and had themselves made a heavy investment in weapons development. With the end of the war, American science became a key force in the rebuilding of schools, laboratories, museums, and technology-based industry throughout much of Europe and in Japan.

Today we face a number of challenges of comparable importance to those that led to a marshalling of scientific expertise in wartime. The nature of those challenges points up very starkly the limitations of traditional policy-making procedures. Issues have become increasingly technical, requiring specialized knowledge for their solution. At the same time, they have become more complex, involving greater unpredictability and calling increasingly for more interdisciplinary approaches. The increased pace of events in the world today demands quicker responses and better information in decision-making. Finally, and most importantly, there is the increasingly global nature of the issues we face. In areas such as the environment, health, energy, telecommunications, education, science policy and the protection of intellectual property, it no longer makes sense to construe problems purely in national terms. In a world that is becoming daily more interdependent, policy-making must assume an increasingly global dimension.

Taken together, these factors demand a new approach in which the scientific virtues of openness and internationalism play a greater role in decision-making. At present, the pattern is still largely a traditional one in which experts in their various fields offer advice in confidence to decision-makers whose interests and reflexes are often more political than scientific.

It is increasingly apparent that this model will no longer serve. Through its increasingly diverse and rapid applications, science has come to impinge more and more on society – vindicating to some extent a faith in the scientific enterprise, but giving rise to doubts about its ultimate viability or about human capacity for the wise management of its applications.

The effect of these changes has been to render obsolete the old model of a two-actor relationship, with scientists functioning as an autonomous 'Republic of Science' while government makes policy decisions in the name of society as a whole. Increasingly, science has had to contend with and respond to many different and not always convergent interests in society. It has had to enter into a many-sided relationship with many different groups in society, including industry, government departments not directly concerned with research or education (for example, energy, environment, culture, foreign affairs, defence) and the public as consumers and tax-payers. This has both complicated and deepened the rivalries within the scientific and technological community itself. The traditional boundary between scientific practice and science policy has become blurred: the issues, though still scientific in nature, are also economic, cultural and political.

The picture is further complicated by the emergence of new issues arising from the kind of questions science is now addressing, and partly from the increasing diversity and impact of the applications of science. These issues – concerning, for example, the earth's environment, risk management in an industrial age and biotechnology – raise fundamental ethical and other questions that lie outside the strict bounds of competence of the scientist.

Yet despite this growing diversification and complexity of actors and issues (and despite a better understanding in recent years of the sociological and cultural conditions governing scientific and technological production), the articulation between science and government continues to be based in large measure on early post-war conceptions. In this as in most other facets of human organization, reflection lags behind events and structures are not attuned to realities and needs. Although the context in which scientific knowledge is produced has significantly changed, almost all governments are ill-prepared to cope with the new situation.

The problem is compounded by the fact that existing science-and-technology policy-making mechanisms are quite inadequate in many countries of the world, both developed and developing. Responsibility for science and technology policy is often sectorally divided, which is not conducive to the development of an integrated policy or to the proper promotion of research in frontier areas cutting across traditional fields or located in their interstices. Where a ministry for science and technology or an equivalent body exists, it is often not sufficiently separate from daily politics to maintain the long-term perspective required in this sphere. In general, there is a tendency for the interests of science and technology to be sacrificed to expediency – for example, to short-term economic considerations or the government's need to achieve immediate results in order to stay in power.

Matters are not helped by the fact that science invariably speaks with many voices. Despite their commitment to objectivity, technical advisers rarely agree once they stray outside the realm of hard science. 'When issues related to science and technology having strong political content . . .

come to the fore,' Frederick Seitz, a past president of the American National Academy of Sciences has said, 'it is always relatively easy to find individuals with good professional qualifications to argue, even vehemently, in support of either side.' This accords undue power to decision-makers who then make choices based on their interests and/or biases, using so-called scientific evidence in support of what is in effect a foregone conclusion.

The lack of agreement among scientists reflects, of course, the complexity of the issues involved. These do not come in neat disciplinary packages labelled scientific, economic, political and philosophical: they invariably involve a complex combination of fields and require a holistic approach for their solution. Scientists by themselves cannot be expected to provide a complete answer to such questions. However, it remains essential that decisions, whether on specific or broad policy issues, should embody the best and 'hardest' possible scientific advice. At the same time, it is important that the spectrum of expert advice should be extended to ensure inputs to the decision-making process from all relevant fields. Even where motivated by the best will in the world, emotions and feelings – often based on stereotypes – have a limited place in this process, which should be based as far as possible on scientific fact and expert opinion.

The foregoing seems to me to highlight the need to take a fresh look at science-and-technology planning with a view to arriving at a relationship between science and government capable of measuring up to present-day complexities. What is required is to move away from what has been termed 'decisionist action' – whereby government decrees policy based almost exclusively on expertise provided in confidence by the scientific and technological community – to an approach whereby government effectively takes into consideration the views of the different protagonists (including certain government departments as well as the scientific and technological community) with their diverse values and standpoints. In this model, government would decide on policy through a process of conciliation and synthesis. This would imply free and unconstrained discussion between expert representatives of the various groups concerned by a specific policy issue. Such an approach would provide a broader base for decision-making and should, at the same time, make for greater openness in science-and-technology policy-making.

The institution of such a system would need to be accompanied by other measures. It presupposes improved access by all concerned to the information on which discussion of the issues and, ultimately, decision-making will be based. In this connection, consideration should be given to the construction of new instruments – such as computerized science-and-technology indicators – to assist in arriving at informed diagnoses. An effort should also be made to raise the general level of scientific and technological culture (or 'literacy') to optimize societal participation and support, an issue to which we shall return.

A new articulation of the kind suggested would also require some fundamental changes of attitude on the part of scientists and politicians alike. Scientists would have to be prepared to relinquish some of their traditional hegemony in the sphere of policy advice. At the same time, they would have to develop a greater sensitivity to the social and political impact of the advice proffered. Politicians, for their part, would have effectively to cede some of their decision-making power to the expert protagonists, while learning to accept the inevitable quota of uncertainty and instability in the advice received.

The need then, as I see it, is to redefine the role of expert advice in such a way as to emphasize the importance of scientifically based decision-making while excluding any idealized vision of enlightenment through scientific rationality. The goal should be to arrive, through the marshalling and confrontation of advice in all relevant fields, at the most robust policy and regulations possible. In this connection, it must of course be borne in mind that timeliness is an essential factor in policy-making. The diagnosis on which policy is based should be as complete as possible, but it must also be made in time: if we wait too long, 'the patient may die' or irreversible damage may be caused. Irreversibility is a key criterion to be taken into account for timely action. Postponement could lead to a point of no return. It is a question of the 'ethics of time'.

In the broader context of policy implementation, thought also needs to be given to the increasing unwieldiness of scientific and technological institutions. Even in the simpler two-protagonist system, the huge size of man's scientific and technological institutions, with their multitude of decision-making centres, does not make for easy control. Consideration thus needs to be given to trimming them down to manageable proportions so that responsibilities can be more readily pinpointed and the risks of duplication lessened. That being said, the hubris of overplanning must be avoided. Scientific creativity cannot be regulated and publicly accountable in every detail. Research is not a linear process. It is 'to see what others cannot see and to think what nobody else has thought', as Professor Hans Krebs once told me in Oxford. Policy decisions in certain areas of fundamental research must rest with the scientists concerned – such autonomy being the very condition of those qualitative advances in understanding that are ultimately as crucial to the applied as they are to the basic sciences.

This is perhaps the point at which to underline – addressing a preoccupation that perfuses the contemporary relationship between science and society – that knowledge itself is always positive. The problem arises with its application, which can be negative or even perverse. It is at the level of applied research, then, that the question of social impact must be considered. Decision-makers must be prepared to promote basic research for the very good reason that, in the words of Bernardo Houssay, 'there are no applied sciences if there is no science to apply'. And it must be applied, as Bertold Brecht has forcefully said, to improve the quality of life, and to reduce suffering. Applied without human targeting, as Franco Ferrarotti has said, 'technology is perfection without purpose'.

The relationship between science and government is not, of course, simply a national matter – there is an increasing need for a world approach to science policy. Many of the problems, as we have already seen, are international in scope, but the need for a world approach follows equally from the existence of a large and growing gap in the scientific and technological capabilities of the developed and the developing worlds. The fruits of science are still very unevenly distributed. UNESCO, which among the United Nations agencies has the leading role in this field, has been trying for many years to narrow this gap through international co-operation. It can take a variety of forms: assisting Member States in the formulation of strategies for science-and-technology development; providing training, research and information exchange in the area of science-and-technology policy; contributing to the popularization of science; awarding of prizes for outstanding research and popularization work; publishing every two years the *World Science Report*; promoting a rapid transfer of knowledge through scientific networks (in biotechnology, molecular and cell biology, brain research, and so on) and through UNESCO's chairs; favouring international co-operation, through important scientific associations such as ICSU, and facilitating interchange of experts through world programmes such as the International Hydrological Programme, the oceanographic programme, the solar and renewable energies programme, the environmental programme (MAB), the microbiological programme (MARCEN), the human-genome programme, and so on.

The role of freedom

During the years of the Cold War, the value of science and technology has tended to be construed in terms of its military and economic applications. While this has enabled research in many specialities to enjoy a golden age of generous funding in nations which saw their survival in terms of scientific expertise, it has tended to widen the gap between the rich and the poor nations. The ending of this chapter of history provides a chance of basing civilization on peace, not on war. The question is whether we are as ready to pay the price of peace as we were to pay the price of war.

The close co-operation between science and government that is needed can flourish only in free societies, so it is encouraging to observe the growth of democracy in recent years in all parts of the world. The twentieth century has so far shown us both the apogee of the state in various totalitarian forms, and its demise as an acceptable answer to our problems. It was Erich Fromm who described totalitarianism as 'the escape from freedom', by which he meant that the twentieth-century state often reflected mankind's fears of having to take responsibility for its own destiny. Instead, it offered the nation the security of leaders claiming to represent the inevitability of history, and the best and most progressive elements in national tradition. By expressing a fear of responsibility for the future, totalitarianism promised – in exchange for the surrender of personal freedom – a guaranteed utopia that would justify the sufferings of the present. In the march towards this utopia, science and technology were given a vanguard role. Yet experience has shown that scientific knowledge cannot be so easily bought and sold.

For a while, and so long as technology remained predictable, centrally planned economies demonstrated a brutal ability to ape the successes achieved elsewhere, and even exceed them. Many Western intellectuals who visited the Soviet Union in the 1930s compared the purposive activity there with the apathy of the democracies, and made no secret of their preference. In a world dominated by coal and steel, the command economy did achieve some successes, at the cost of great human misery. But when the focus of development shifted to industries based on new knowledge in electronics, agriculture, biology and biotechnology, the price of suppressing free thought and action had to be paid. The centrally planned economies, for all their claims to be 'scientific', lost the battle because they used science as an instrument, obscuring its true nature. The lesson is that nobody, however powerful or well informed, can know for certain which scientific ideas will shape the world of tomorrow. Only by allowing scientists the freedom of action to go where their curiosity leads can we be sure that science will flourish and our economies will flourish with it.

Of course, the problems faced by societies in transition from oppression cannot be solved simply by appeals to the free market, effective as it has so often proved. The reforms undertaken in Eastern Europe, as well as in certain Latin American or African countries, seem to place an excessive hope in a rapid transition to a free-market model. Undertaken in a context of considerable indebtedness, weak planning institutions and uncertain currencies and markets, the management of such a complex transition is no easy matter, particularly when the historical memory of the 'culture of freedom' is lacking. The free market itself – the so-called Invisible Hand – is no guarantee of meeting long-term needs or of ensuring the best use of human resources. The question is whether we possess adequate knowledge to devise sound methods of moving from centrally planned economies to a freer but none the less humane market. A longer-term view, based on investing now to prevent future catastrophes, is required, both for practical and ethical reasons. This is in my view most important, for those who have been suffering under the shadow of oppression or extreme poverty will be very disappointed if we offer them only the rules of the free market.

The sharing of knowledge: science as a mass phenomenon

The success of science is based upon shared knowledge. Only by open publication of results and the mutual testing of ideas can we be sure that we are moving along the right lines. Individual scientists of genius may have the courage to go their own way, but they will achieve nothing unless their theories take into account what others elsewhere are doing. The extent to which knowledge is shared, both within nations and between nations, will determine our ultimate success.

As scientists, many of us are painfully aware that we deal, in our laboratories and classrooms, and even in our role as public servants, with realities and constructions of reality that go beyond the everyday 'common sense' about nature that subsists in everyday thought and action. Scientists and those who work closely with them inhabit an intense and dynamic intellectual community which has already internalized the most recent breakthroughs in scientific knowledge as part of our road maps to everyday reality. Yet the community at large, all too often including its leaders, remains years behind in its understanding of natural phenomena and of the directions in which science is moving.

Scientific illiteracy is one of the key issues we face. Elected leaders have neither the time nor the resources to grasp biological or environmental complexity, although they may be required to make budgetary and policy decisions that will determine the educational, scientific and ecological priorities for years to come. Voters are unable to make rational decisions on issues that require basic scientific knowledge, because all too often they have never been provided with that knowledge. The problem is more than one of scientific literacy or, to put it more politely, the absence of a scientific culture within our political culture. An unprepared community cannot truly govern itself nor can it plan for its children's future.

We live in a society that depends on science and technology without really understanding it. In January 1992 *The Times Higher Educational Supplement* ran a front-page article quoting the views of 'leading academics and writers' in the United Kingdom on the choice of ten 'essential texts for the educated person'. One looked in vain for anything scientific in this list. There was no mention of Charles Darwin or Albert Einstein or Neils Bohr; to be a truly educated person by this definition does not require one, apparently, to have read any science at all. As the geneticist Richard Dawkins put it at the time: 'Educated is synonymous with being educated in the humanities.'

Surveys tell us the same story. When basic scientific knowledge was tested in one survey, it was found that less than a third of Britons, and 43 per cent of Americans, knew that electrons were smaller than atoms. Almost a third of the British respondents, and a quarter of the American ones, believed that antibiotics kill viruses as well as bacteria. I am not seeking here to be critical of any particular nation: there is no reason to suppose that these results would be any different if the surveys had been conducted elsewhere.

Nor is the antipathy and ignorance of science and technology anything new. We tend to think of the nineteenth century as a period of confident expansion, ruled by engineers. But even then a strong undercurrent of opposition thrived. It is easy to laugh at the Duke of Wellington, who opposed the development of railways because, he said, 'they would enable the lower orders to go uselessly wandering about the country'. But he wasn't alone. The French novelist Gustave Flaubert listed the four greatest misdeeds of modern civilization as 'railways, factories, chemists and mathematicians'. Yet he happily used the railway as a convenient way of meeting his friend Louise Colet, who lived in Paris while he lived near Rouen. They would meet half-way, at Mantes, a two-hour journey for both of them instead of a dusty day in a horse-drawn carriage. One might have expected Flaubert to be grateful for the train, but he wasn't.

Many of us harbour the same double standard: we enjoy the benefits of technology while remaining largely ignorant of its inner workings. We have a society where, in Martin Weiner's words, we have the outward acceptance of modernity without the inner conviction. Few people know, or very much care, what goes on under the bonnets of their cars, or at the other end of the high-tension cables that supply their houses with electricity. To them it is all a bit of a conjuring trick.

Ignorance is forgivable, and indeed inevitable: we cannot know everything. But to claim that the world was a better place when such ignorance was universal is perverse. John Ruskin, the nineteenth-century moralist and aesthete, helped establish an anti-industrial view even as the great engineers were transforming the world for the better. He argued that the rise of industry had in no way changed the fundamental truths of life.

> To watch the corn grow, and the blossom set; to draw hard breath over ploughshare or spade; to read, to think, to have, to hope, to pray. . . . The world's prosperity depends upon our knowing and teaching these few things; but upon iron, or glass, or electricity, or steam, in no wise.

The language is wonderful, but the sentiment nonsense.

Ruskin painted the world as a choice: between stability or restless change, between pastoral contemplation or the industrial inferno. He left no doubt where his own sympathies lay. His values are unconsciously echoed by many people today, from environmentalists to animal-rights activists and those who oppose genetic experiments in the name of the sanctity of man. The common characteristic of such groups is their assumption of the moral high ground. Against such foes, science needs to be constantly alert. There is nothing intrinsically immoral in technology or in industry: Ruskin's heirs must not be allowed to dictate the terms of the debate.

It is clearly impossible to provide every person with a profound understanding of all the important and controversial areas of science and technology; impossible because even scientists themselves do not possess it. A cell biologist does not understand particle physics; a metallurgist is unlikely to have more than a limited grasp of genetics; and hardly anybody understands the wonderful world of higher mathematics. What singles out scientists and engineers is not their knowledge of fields other than their own, but their conviction that they could understand these other subjects if they wanted to; a confidence that is not generally shared by non-scientists.

This is the confidence we must try to spread more widely. The evidence is that when scientific ideas address people's concerns and interests, they show an impressive ability to locate information and translate it into forms they find useful. An interesting example of this is the annual AIDS conferences, ten of which have now been held since the disease first challenged the scientific community to find an answer in the mid-1980s. While ostensibly gatherings to discuss research in progress, the conferences also involve many people who are HIV-positive, or who care for AIDS sufferers. These people are not scientists, but because of the immediacy of the issues for them, they rapidly become familiar with all the latest research in the fields of virology and immunology, two of the most difficult of scientific disciplines. AIDS activists make life very difficult for scientists, with their sometimes unreasonable demands for faster progress. But they also make the whole field more stimulating, and contribute towards ensuring that research is focused on the real problems.

The experience of AIDS exemplifies what academic studies have told us about the acquisition of scientific knowledge. In the view of Dr Bryan Wynne of the University of Lancaster: 'The public uptake of science is not based upon intellectual capability as much as socio-institutional factors having to do with social access, trust, and negotiation as opposed to imposed authority.'

This gap between what scientists know and what the public understands is one of the greatest threats to clear-sighted science and public policy. Speaking in 1931 to an audience at CalTech, Albert Einstein said:

> Concern for man himself and his fate must always form the chief interest for all technical endeavour . . . in order that the creations of our minds shall be a blessing and not a curse to mankind. Never forget this in the midst of your diagrams and equations.

On the scientific side we in the laboratories and classrooms must share in the blame for the lack of diffusion of our findings and concerns through the communities in which we live. On the other side, however, the political leadership – local, national, and international – must also bear the responsibility of leaving science in a ghetto of narrow specialisms, despite its central importance in the world around us. The responsibility for scientific illiteracy is shared and the solution to a continued state of chronic ignorance or misplaced sensationalism must be found at the meeting point between scientists and community leaders.

There are no easy answers, but it is clear that scientific literacy must try to provide at least three things.

First, a whole picture of the different issues; a view from different angles; by oversimplifying complexity we can produce false perceptions which lead to wrong decisions. Only a holistic approach can help us evaluate, in all their complexity, the different options open to us. One example is that of energy sources: the costs and risks of a power station are evaluated by most economists as if coal were a renewable resource, picked from the tree like an orange. Coal, however, is not renewable. Other examples are these: can we seriously address the problems of developing countries without taking into account the terms of an equitable international trade? Or can we try to improve the education of rural women without providing the villages with wells and drills for obtaining a pure water supply?

Secondly, it also means a far-sighted vision, bearing in mind that in all the most important questions the most fragile and ephemeral factor – biologically, at least – is the decision-maker. Mortality is part of human existence, but we must learn to think ahead for the intergenerational human rights of our children and their children.

Thirdly, we need a historical and global perspective, in order not to forget that there is a past which has forged our own behaviour and which, when consciously understood, can be used to change our ways of being. Nothing is determined by some mechanical 'logic' of history: everything depends on what we do in the here and now. This is the legacy we can give to our descendants.

Is it utopian to think seriously about making science part of popular culture? I think not. UNESCO has recently concentrated both programme resources and budget on such fundamental questions as Basic Education for All, and Environmental Research and Environmental Education, steps which I believe will help place science at the core of everyone's culture. Moreover, I believe that UNESCO's strengthened focus on partnerships with the professional communities is a way of bridging the gap between the political and the scientific. Ultimately, we cannot conceive of a harmonious and stable world without scientists of the very highest calibre working and teaching even in the poorest settings. Literacy and scientific literacy would go a long way towards solving many of the problems of the poorest countries.

Science in the mass media

Forty years ago the novelist C. P. Snow declared that we live in a world of two cultures. One, the artistic, is generously reported in newspapers and on radio and television, while the other, the scientific, must make do with much less space. Why is one so lavishly covered and the other so frequently sidelined? It is dangerous to generalize, for some countries have a press that is more open to scientific articles than others, but in general it must be true that science occupies fewer column inches than its importance merits. On television, the disproportion is even greater.

There are some real reasons why science coverage is never going to match that of the arts. The bulk of arts coverage is not news, but reviews. Science is not a performing art, to be reviewed by critics, nor is it a spectator sport. It is about ideas, and new and influential ideas are always going to be scarce, in art as well as in science. If opera critics were obliged to write only about new operas, and never about revivals of old ones, they would have a thin time; yet this is what science writers are expected to do. They cannot simply spin tales about the same few ideas and personalities that provide the political writers with their raw material. A science story is expected to contain some kernel of new information – and despite the huge amount of science that is done, genuinely interesting new ideas are rather rare.

Having said that, however, there is clearly a need to improve the dissemination of science in the mass media. But to do so involves some understanding of the dynamics of news-gathering, and the way in which a newspaper or a news broadcast is assembled. Many scientific developments may be interesting, but they will not be widely reported unless they can satisfy a few simple tests. These include: How many people is this going to affect? Will it change anything? Will anybody be infuriated or outraged by it? Does it have implications for government policy? Will it be, in that magic phrase, 'a talking point' – something people will discuss at home or at work? Is it so outrageously improbable that it scores simply on grounds of implausibility? There are many reasons for deciding that one story meets these criteria and another does not, which working journalists learn by experience. These reasons often seem strange to those outside the media. A particular failing is that newspapers have a short attention span, and quickly lose interest in scientific issues that go on and on without ever being resolved. Since these include many pressing problems such as environmental degradation and global warming, there is a danger that the public is being short-changed. But newspapers require answers, not judicious fence-sitting exercises.

The explosive growth of new media of communication offers many opportunities. Communication, information and computer science are merging to create new formats – such as multimedia – and new methods of access to information through telecommunications. Satellite and cable have increased the number of television channels available, and have also enriched the mixture. The satellite Discovery Channel, for example, which is now widely available, is devoted to programmes about science, technology and the natural world.

The information revolution is going to offer some extraordinary opportunities in the transmission of knowledge, and in the techniques of distance and non-formal learning. To reach its full potential, however, the expansion of the media must be matched by a comparable increase in its variety and quality. For every Discovery Channel there are likely to be many more which offer the same mixture of game shows, situation comedies and soap operas with which we are already familiar. Of course, popular entertainment is not to be despised, but the power of the medium is such that it would be a waste if that was all it was able to provide.

The increasingly global nature of the media also imposes new obligations. Producers, publishers, operators and broadcasters must be aware of their ethical obligations, for their output disregards frontiers. As a result, most of the regulations that have traditionally governed broadcasting are becoming increasingly unenforceable. In the world of the new media, professional ethics and better training are going to have to take the place of government rules.

The role of the scientific community

Un día / los intelectuales / apolíticos/ de mi país / serán interpelados / por el hombre / sencillo / de nuestro pueblo./ Se les preguntará / sobre lo que hicieron / cuando la patria se apagaba / lentamente / como una hoguera / dulce, pequeña y sola[2] (Otto René Castillo).

Science is both a culture and a tool. It provides illumination both literally and metaphorically; and in general where science is in shortest supply, there one will find the greatest human needs. It is among the developing nations, for example, where the lack of scientific literacy is most acutely felt. If we are to solve the whole complex of problems, of which environmental degradation is one, we have to think in terms of developing human potential worldwide. In different ways and in different degrees, the same challenges applies to North, South, East and West – to the developing countries where damage can still be prevented and to the industrialized world which is the source of so many of the problems. But the problem is posed most acutely in the developing countries.

There is little need for me to repeat the gloomy statistics that summarize the plight of almost three-quarters of the world's population. But a few figures may help to put the problem in perspective. Enrolment in higher education in the developing countries as a whole amounts to only 8 to 9 per cent of the age-group between 18 and 23. In the least-developed countries, the figure is a mere 2 per cent. This is to be compared with an average of 37 per cent in the industrialized countries. Scientific and technical personnel amount to less than 1 per cent in the developing world, against more than 8 per cent in the industrialized world. Despite having 80 per cent of the world's population, developing countries are

2. One day, the apolitical intellectuals of my country will be called to task by the ordinary man of our people. He will ask them about what they were doing while the nation was slowly dying out, like a bonfire, gently, small and alone.

responsible for only 4 per cent of the world's research-and-development expenditure, and they possess only 5 per cent of the world's computers.

These figures are disturbing because they concern sectors vital to growth in a knowledge-intensive world. Gaps in these areas tend to be self-reinforcing, backed as they are by the loss of many talented people from developing countries through the brain drain, and the increasing privatization of basic scientific and technological information.

Ultimately, each country's capacity to move towards sustainable development will depend on the supply of a critical mass of skilled manpower in the many scientific and technical fields needed. A country's capacity to promote development depends also on appropriate policy, legal, regulatory and political frameworks. It depends on its capacity to choose the appropriate technology. Skills, knowledge, technical know-how and a sufficient number of trained people are necessary for developing sound endogenous technologies and for absorbing and adapting imported technologies through technology transfer.

The training of the right kind and number of scientists, experts, technicians and teachers is a critical need, perhaps the most critical of all, especially in Africa. UNESCO's work in this field has identified a number of key problems.

The first is the lack of a 'critical mass' at national level. Despite considerable efforts, many developing nations do not yet have enough trained people. In some countries, notably in Africa, there is still need to build up a corps of first-generation scientists and engineers able to train the next generation. In many countries there is a shortage of training facilities, especially those relevant to preparing teams of specialists to conduct studies and research useful for decision-making.

A second problem is the lack of relevance of established institutions and programmes to the local conditions. Too often, specialists in developing countries have been trained abroad, where conditions are very different. This can encourage the introduction of alien models of production, management and consumption. Curricula often fail to address the needs of their societies, and teachers are insufficiently aware of local problems. As a result, many developing-country trainees receive an education that has little bearing on their country's social and economic needs. Updating is essential. The universities must become centres for intensive, lifelong education, with the aim of creating high-qualified citizens. Everything can be expected of those who have knowledge and skills; without such skills, development will not take place. In the coming years, a lot of imagination will be needed to ameliorate the quality of life in the rural areas, providing electricity to those areas now lacking it, avoiding migration to the urban slums. One possibility is the use of military forces during times of peace to become 'friendly development partners', as a position paper for the Social Development Summit of June 1994 put it, or to stimulate the building of housing and new urban transport systems as an alternative to weapons.

A third problem is lack of continuity. Students returning home with skills inappropriate to local needs often end up abandoning their professional fields for other pursuits, such as business or government. They may still make an important contribution, but their direct input to science and technology is lost. The lack of appropriate institutions also results in a disruption of research and of development programmes oriented towards national needs. Here is where the absence of an indigenous talent pool, the 'critical mass', is most acutely felt.

A fourth problem is shared by developed and developing countries alike. Almost all current training programmes are sectoral or disciplinary, and do not address the complex interactions between people, resources, environment and development. This is a major challenge in training the next generation of scientists, who will need to have the appropriate interdisciplinary insights to tackle these problems.

The challenge of biotechnology

Let me try to illustrate both the problems and the possibilities by considering the most rapidly growing field in science today, biotechnology. The explosive growth of new ideas and techniques in molecular biology and genetics are likely in the next generation to have widespread applications in pharmacy, medicine, veterinary science, agriculture, food and nutrition, and other fields. The potential for developing countries in improving health, agriculture and nutrition are clear, but so is the danger: that the biotechnology revolution will lead to a closing of the loop of knowledge, as governments and corporations protect themselves by seeking to patent important discoveries. There is no doubt that the poorer countries, who have perhaps the most to gain from biotechnology, would be the losers in any such scientific 'trade war'.

The significance of biotechnology for the Third World can be illustrated by the example of agriculture in Asia. By the early 1990s, Asia had more than 50 per cent of the world's population, over 70 per cent of the world's farming families, but only 25 per cent of the arable land. By the beginning of the twenty-first century, the area of land available will be 0.1 hectare per head in China and 0.14 hectare in India. The Asian population growth rate is 1.86 per cent. The only way for countries such as China and India to feed their populations is to achieve continuous improvements in yields. China has engaged in large-scale exploitation of hybrids for this purpose.

Since 1950, the areas of irrigated land in the world has expanded threefold, with much of the expansion occurring in the developing countries. The scope for further expansion is more limited. Improved water management, and the incorporation of genes for drought-tolerance are therefore urgent necessities. In livestock husbandry, too, the needs of the developing countries could be met in part by the use of biotechnologies pertaining to genetic improvement, veterinary medicine and nutrition.

Biotechnology is expected to offer answers to a range of problems relevant to the Third World. In agriculture, it could provide maize resistant to the stalk-borer, virus-resistant cassava, protein-enriched cassava, cutworm-resistant potato, and field trials for diagnosing fungal diseases. Some of these are available now, and others are expected soon. Nitrogenase capacity – for the direct uptake of atmospheric nitrogen by plants – is one of the most important objectives in plant genetics, as it will allow reduction of the use of fertilizers. In UNESCO, the Biotechnology Action Committee promotes, through scholarships, networking, and intensive courses, the speedy training and updating of scholars from the developing countries. In human health, a hepatitis-B vaccine is available now, a schistosomiasis vaccine should be available soon, and a vaccine against malaria remains a longer-term goal.

These are major goals of huge value to the Third World, but there are some concerns. Can the 'biotechnology revolution', organized as it is largely by companies based in the industrialized countries, really help poor farmers in developing countries? Several factors seem to argue against it. The bulk of the effort is likely to be directed at crops of interest to industrialized farmers of the North, and even when the results are also beneficial to tropical agriculture, there may be a high price to pay. The extension of intellectual property rights to individual genes or genotypes could have an impact on the availability of such improved material to poor farmers. They might also be prevented from saving their seed for the following year's crop. Would intellectual property rights be exclusively confined to rewarding formal innovation, even though the informal innovation system has played, and is still playing, a key role in the conservation of plant and animal genetic resources? Another worry is that research might be directed at finding substitutes to natural products, such as the high-fructose corn syrups and other natural or synthetic sweeteners used as a substitute for cane sugar, as well as substitutes for vanilla flavour and cocoa butter.

Some of the fears may prove groundless. The increasing globalization of the industry has reduced local and national pressures for substitution which were, in any event, driven less by cost than by health and nutrition arguments. But it is clear from even a brief consideration of this new technological opportunity that the Third World cannot afford to be a passive onlooker while all the progress is made by the developed countries. Among other important reasons is global security, because the expansion of markets requires a growing number of countries able to be part of the trade cycle. Dependence on the knowledge, technologies and cultural goods produced by a few major countries is a fact; but nobody can be content to be a spectator of scientific development for ever. Peoples wishing to decide their own future will have to contribute, however modestly, to the global stock of knowledge and will have to be able to share in it. In particular, research on the major staple food crops in the developing countries needs to be supported internationally, for the sake of global sustainable development. The same holds good for diseases prevailing in developing countries, which require an international effort.

The transfer of knowledge

To ensure that the gap between rich and poor does not widen further, we need new ways of transferring knowledge from those rich in it to those who are far more in need. These should include networks and databases, easily accessible to specialists in the Third World, who often suffer an extreme intellectual isolation. They should also include intensive short-term scholarships at the best centres of research in the richer countries, as well as sending top specialists out to the Third World to develop talent *in situ*. The poor countries themselves might do well to abandon imported educational models already obsolete in their countries of origin, in favour of less rigid systems which can produce large savings. As well as improving the situation, such reforms can reduce the problem of the brain drain, which needlessly redistributes scientific and other talent away from where it is needed to where it merely happens to be better paid and better equipped.

UNESCO launched a plan of action called UNITWIN, whose purpose is to bring about closer co-operation between universities at a time when reform of higher education is everywhere both necessary and urgent. The idea is to foster a spirit of solidarity based on twinning arrangements, networks and other forms of inter-university co-operation throughout the world, including North–South, East–West and South–South co-operation. A system of 'UNESCO chairs' established under this initiative aims to increase the teaching and research capacity of higher education establishments in specific disciplines, mainly in the poorer countries. It gives prominent specialists the chance to teach as visiting professors in these countries, promoting the transfer and sharing of knowledge. The organization is also promoting two initiatives in Uruguay and Venezuela: the *enlace* (link project) and the MECCO (Mercado Común de Conocimiento, or Common Market of Knowledge). Both will facilitate the interrelationship between countries, universities and research centres.

The responsibility of scientists

> Nada se consigue de un golpe, ni surge por sí mismo, ni está seguro para siempre. El hombre se inventa y se crea en la incertidumbre y el riesgo[3] (Jacques Spaey in *El desarrollo por la ciencia*, UNESCO, 1970).

Knowledge is always good: it is its applications that can be harmful or even evil. A hammer can be used improperly as an instrument of aggression; so at another level can atomic energy. But we must be careful not to allow only the negative aspects of science to be emphasized. These are usually described in exaggeratedly dark colours, while the good aspects are not news. The scientific and technological community has a duty to remind people of the enormous benefits science has brought. Some years

3. Nothing is achieved at a stroke, nor rises without reason, nor is certain for ever. Man invents and creates himself amidst uncertainty and risk.

ago, the Green Revolution, which has transformed the future prospects of millions of people in Asia, came under criticism from environmentalists because it had tended to favour richer farmers, who could afford the pesticides and fertilizers on which the new varieties of wheat and rice depended. The critics behaved as if this discredited the entire enterprise, which was far from the truth. They were equating the avoidance of starvation for millions with an effect, doubtless real, but many orders of magnitude less important.

There are plenty of other examples. We need to sell our wares, and there could be few better consumers than today's. Does anybody think about poliomyelitis any more? Or of the way we have been able to reduce pain and even eliminate it with the help of analgesics or modern anaesthetics? Or of the transport and telecommunications systems that enable us to be citizens and witnesses of the whole world, even though they compel us to assume the responsibilities this entails?

Nevertheless, we cannot shirk our moral responsibilities. We must admit that there have been a considerable number of silent scientists in the twentieth century. The possession of knowledge carries with it a moral obligation. What would we think of a meteorologist who did not warn us that a storm was coming? Or of clinical analysts who failed to alert us to the urgent nature of their findings? Scientists have to transcend the walls of their laboratories, their departments, their faculties, and their industries. To say that the scientific life is simply a matter of 'publishing or perishing' is both disparaging and wrong. A scientist who chooses to remain silent, particularly when irreversible consequences may follow, is behaving both unprofessionally and immorally.

In no field of science are the issues more complex than in genetics, nor is there any other area of science in which the discriminatory use of the information would be more scandalous. The whole subjects of bioethics, far from being a luxury, should be the preserve of the entire human community, since the advances made by the few are a battle won for the whole human race. Bioethics came into being, rightly, at the moment when techniques of genetic manipulation first emerged in the 1970s, and questions began to be asked about the influence of molecular biology on the future of humanity. Far from simply being concerned with an investigation of the relations between science and society, it is concerned with the relation between man and nature in its biological diversity. The development of bioethics illustrates, I believe, a proper and appropriate response by scientists and philosophers to the social questions raised by science.

These are many. Genetic diagnosis and screening make it possible to give reassurance to members of families with a history of genetically transmitted illnesses. However, should these tests be conducted only after the disease appears? Should genetic testing be restricted to hereditary, incurable, fatal or incapacitating diseases? How can the confidentiality of information be guaranteed over time, especially when they could be of enormous interest to employers and insurance companies? For the time being, the scientific community has agreed that there should be no use of gene therapy on the germ cells – those that would confer the changes to the next generation – and that the cloning of human beings should be banned. But is this agreement based on ethical principles, or does it merely indicate that everyone at present considers these techniques to be untimely in view of the current state of knowledge and the possible associated risks?

Behind the first series of questions looms the spectre of eugenic selection, with its attendant host of violations of human rights and individual freedom. The temptation for couples to use the results of screening in selecting their potential offspring is understandable, but it is difficult to draw the line between the legitimate desire not to pass on serious genetic diseases and the illusory quest for the 'perfect child'. At another level, it is not impossible to envisage entire societies being tempted to practise eugenic selection on entire populations that it might consider genetically vulnerable or as presenting 'risks'.

Each society must work out for itself a route through this moral maze, but there is also a role for the international community. National rules may vary from country to country, which could encourage a form of 'genetic tourism', in which people would be tempted to take advantage of laxer regulations abroad. The first stirrings of this have been seen in the use of techniques of *in vitro* fertilization which allow women past the age of the menopause to become pregnant. In Italy, there are no rules or conventions prohibiting the use of this technique for women in their late fifties or even sixties, so that women from the United Kingdom have travelled there to take advantage of a treatment they would be denied at home because of their age. Without acting as judge and jury in such cases, it is nevertheless clear that some international standards are desirable. Similar arguments apply to the even more vexed question of the patenting of the human genome, where so far different countries have adopted different approaches. For these very reasons, UNESCO has created a World Committee on Bioethics; perhaps it would be wise to extend its scope to include every impact of science on human life.

To summarize, it is clear that the old models of scientific advice, and the old relationship between science and power, are no longer adequate. It was Einstein who said 'the world is one, or none'. Applying this vision of wholeness to the relationship between science and power will undoubtedly call for novel and heterodox solutions. The coming decades will see a progressive 'scientification' of the decision-making process at all levels – particularly at the municipal one ('Science at the Town Hall' would be a good motto). This will take place in the context of an expansion of the advisory framework consistent with a greater recognition of the complexities of the science-policy issues and the need for holistic approaches to their solution. Finally, there will be an increasing need to recognize the global nature of science-policy decisions, and the need to establish a new relationship between science and power that acknowledges that international dimension. Knowledge is power: power to produce, to foresee and to prevent. To apply that knowledge for the benefit of mankind is wisdom. Knowledge and wisdom are the two main pillars of a better, common future.

Contributors to this volume

Raymond DAUDEL

Biologist. He is professor at the Pierre and Marie Curie University, Paris. His works are mainly devoted to applications of molecular sciences to chemistry, biology and medicine. He belongs to various national and international academies and he is President of the European Academy of Arts, Sciences and Humanities. Author of several publications.

Franco FERRAROTTI

Sociologist. Professor at the La Sapienza University, Rome. Founder of the *Quaderni di sociologia* in 1951; at present he is the editor of *La critica sociologica*. Fellow at the Center for the Advanced Study in the Behavioral Sciences at Palo Alto, California, in 1964; Directeur d'Études at the Maison des Sciences de l'Homme in Paris in 1978; he has taught at various universities throughout the world; among his books *Le retour du sacré* (Paris, Klincksieck, 1993) and *The Temptation to Forget* (New York and London, Greenwood, 1994).

Augusto FORTI

Geologist and geophysicist. At present, Senior Special Adviser to the Director-General of UNESCO for Europe and North America. Formerly Director of Research and Training Institutes and Laboratories. He is the author of several publications on scientific research, science and society. Editor and co-director of the review *Prometheus*. Member of the World Academy of Art and Science.

Gérard HUBER

Doctor of psychoanalysis. Délégué Général of the Association Descartes. Author of several publications and television programmes on medical, biological and social subjects.

Alexander KING

Physical chemist. Science adviser to the United Kingdom Government, Director-General of OECD; Co-Founder and President Emeritus of the Club of Rome.

Federico MAYOR

Biochemist. Director-General of UNESCO. Involved in several academic and political activities, he became Rector of the University of Granada. He was Minister for Education and Science in Spain. He is the author of several publications on science and society as well as of many books of poetry.

Ilya PRIGOGINE

Nobel Prize-winner for Chemistry. He is Professor at the University of Texas at Austin and at the Instituts Internationaux de Physique et de

Chimie, Brussels. Author of fundamental research in chemistry and thermodynamics as well as of several books on the relation between science, society and philosophy.